Pathways to Math Literacy

Dave Sobecki

Miami University Hamilton

Brian Mercer

Parkland College

MTH 58/98

Portland Community College

1 2 3 4 5 6 7 8 9 0 SCI SCI 18 17 16 15

ISBN-13: 978-1-259-73116-7
ISBN-10: 1-259-73116-2

Solutions Program Manager: Joyce Berendes
Project Manager: Vanessa Arnold

A Letter from the Authors

To Instructors

"Why do I need to know this?": the bane of the math teacher's existence. Of course, we know that the benefits of mathematical education go far beyond using specific procedures. It's about exercising the brain, learning problem-solving skills, and understanding the importance of being numerate in our society. But what if we really considered that question in a deeper way: **what do non-STEM students really need?** And what if we agreed to move past the "this is important because it's important" mentality, and thought about the topics and activities that will best serve a group of students that are, for the most part, poorly served by traditional developmental algebra?

Our project is the result of attempting to do just that. Most importantly, it's not about watering down the curriculum in an attempt to pass more students. It's about providing non-STEM students with an alternate pathway that will get them into the college-credit math courses they need without getting trapped behind the roadblocks that the traditional developmental math track have become. But more importantly, **it's about focusing on context and critical thinking, and showing these students why the math they've struggled with for so many years is relevant in their lives.**

What we've discovered along the way is how much richer the experience can be for non-STEM students when they stop trying to memorize and mimic, and start to really think and learn. By using a workbook format, focusing on active learning, incorporating technology, and approaching every single topic from an applied standpoint, we've been able to build a course (and a book) that elicits our favorite response from students: **"This doesn't feel like a math course. We're kind of using math… ." YES. Yes we are.**

If you look very carefully, you'll find many of the topics that typically make up the core of the developmental algebra curriculum. We like to think of it as giving your children medicine they don't want by mixing it into a bowl of ice cream. By making everything contextual, and liberally mixing in important study skills and a variety of topics that are usually in the province of liberal arts math, we're making the process of learning useful problem-solving skills through algebra more palatable for students, which at the end of the day is a significant part of the battle. When your students open their minds to the possibility of really understanding a math course, and really seeing how math can be useful, they blossom into the learner that we try to bring out in all of our students.

In some cases, students may be inspired to change their path, moving into a STEM-related field. If so, that's great! Taking the Pathways course in no way precludes that. In fact, the emphasis on conceptual understanding will more than likely make students better equipped to succeed in a further algebra course than if we continue to force-feed them the same algebra that they choked on in previous courses, either at the high school or college level.

To Students

One of the first things you'll notice is that this isn't an algebra course. You will do plenty of algebra, but most of it will be mixed in with doses of interesting topics that require critical thinking and problem-solving skills. We hope you are able to see many places where math seems useful and interesting. Not only can the experience of completing this course help you to be a better student, but it can also help you with life after college. Imagine that!

You may have sat in previous math courses and wondered, "Why do I need to know this?" Good for you! **We encourage you to keep asking that question**. And we hope you find this course to be a rewarding experience that has positive consequences for years to come.

We'd love to hear feedback from instructors and from students. Don't hesitate to contact us!

Dave Sobecki
Miami University Hamilton
davesobecki@gmail.com

Brian Mercer
Parkland College
bmercer@parkland.edu

ALEKS®

McGraw Hill Education

ALEKS is a unique, online program that significantly raises student proficiency and success rates in mathematics, while reducing faculty workload and office-hour lines. ALEKS uses artificial intelligence and adaptive questioning to precisely assess a student's knowledge, and deliver individualized learning tailored to the student's needs. With a comprehensive library of math courses, ALEKS delivers an unparalleled adaptive learning system that has helped millions of students achieve math success. ***NEW*** **Math Literacy pie available.**

ALEKS Delivers a Unique Math Experience:

- **Research-Based, Artificial Intelligence** precisely measures each student's knowledge
- **Individualized Learning** presents the exact topics each student is most **ready to learn**
- **Adaptive, Open-Response Environment** includes comprehensive tutorials and resources
- **Detailed, Automated Reports** track student and class progress toward course mastery
- **Course Management Tools** include textbook integration, custom features, and more

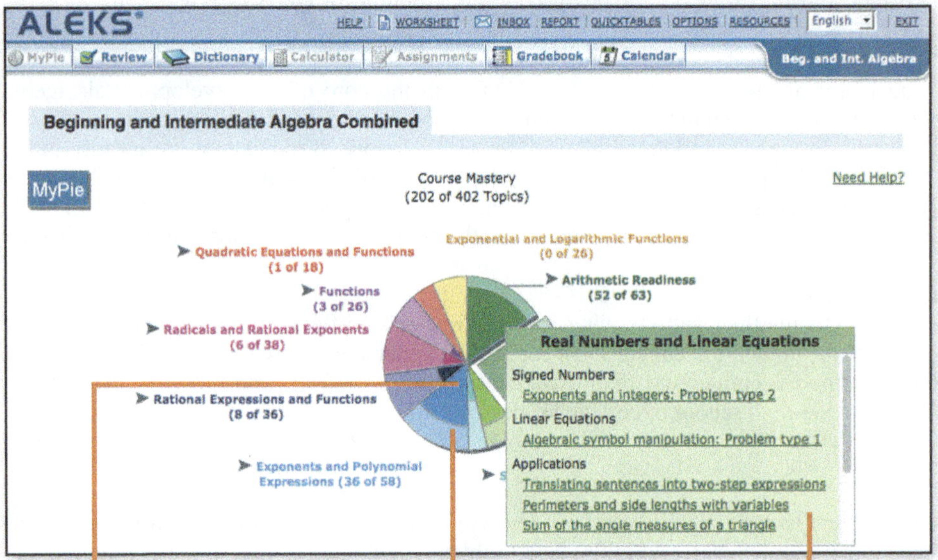

The ALEKS Pie summarizes a student's current knowledge, then delivers an individualized learning path with the exact topics the student is most ready to learn.

Dark portion represents what the student knows.

Light portion represents what the student still has to learn.

Ready to Learn topics appear in pop-up boxes when the student scrolls over a pie slice.

> "My experience with ALEKS has been effective, efficient, and eloquent. **Our students' pass rates improved from 49 percent to 82 percent with ALEKS.** We also saw student retention rates increase by 12% in the next course. Students feel empowered as they guide their own learning through ALEKS."
>
> **—Professor Eden Donahou, *Seminole State College of Florida***

To learn more about ALEKS, please visit: **www.aleks.com/highered/math**

McGraw-Hill Connect® Features:

- Straightforward course and assignment set up
- Lecture capture with Tegrity
- Interactive learning content
- Group Assignments
- At risk reporting
- Customizable content
- Simple integration with every learning management system (LMS)

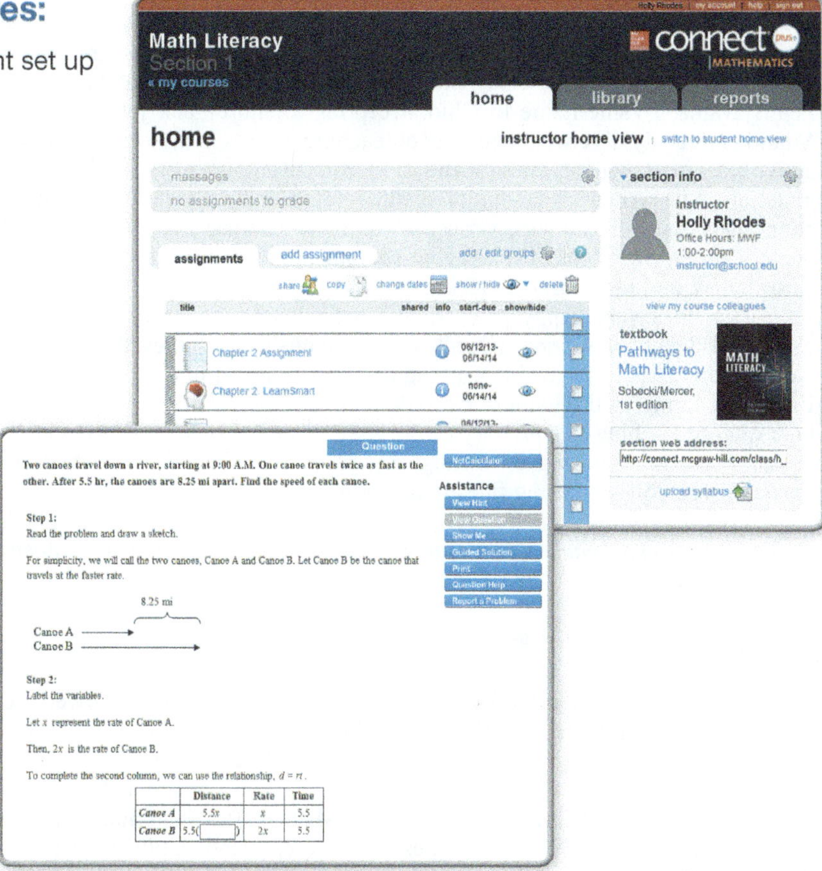

Online exercises and solutions were developed by faculty to provide a seamless transition from textbook to technology.

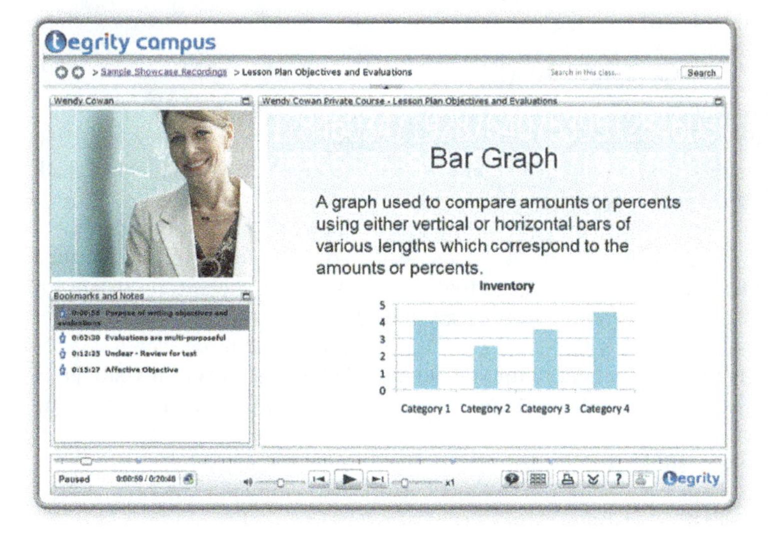

McGraw-Hill Tegrity®

Tegrity is a fully automated lecture capture solution, now available within Connect.

To learn more visit **www.SuccessInMath.com**

About the Authors

Dave Sobecki

I was born and raised in Cleveland, and started college at Bowling Green State University majoring in creative writing. Eleven years later, I walked across the graduation stage to receive a PhD in math, a strange journey indeed. After two years at Franklin and Marshall College in Pennsylvania, I came home to Ohio, accepting a tenure-track job at the Hamilton campus of Miami University. I've won a number of teaching awards in my career, and while maintaining an active teaching schedule, I now spend an inordinate amount of time writing textbooks and course materials. I've written or co-authored either seven or twelve textbooks, depending on how you count them, as well as a wide variety of solutions manuals and interactive CD-ROMS. I've also worked on an awful lot of the digital content that accompanies my texts, including Connect, LearnSmart, and Instructional videos.

I'm in a very happy place right now: my love of teaching meshes perfectly with my childhood dream of writing. (Don't tell my publisher this–they think I spend 20 hours a day working on textbooks–but I'm working on my first novel in the limited spare time that I have.) I'm also a former coordinator of Ohio Project NExT, as I believe very strongly in helping young college instructors focus on high-quality teaching as a primary career goal. I live in Fairfield, Ohio, with my lovely wife Cat and fuzzy dogs Macleod and Tessa. When not teaching or writing, my passions include Ohio State football, Cleveland Indians baseball, heavy metal music, travel, golf, collecting fine art, and home improvement.

Dedication: To my beautiful wife Cat, who lives with far less husband than she deserves while I pursue my dream. This is for us, Shmoop!

Brian Mercer

I can say without a doubt that I was made to be in a classroom. I followed the footsteps of my father, a 35-year middle school math teaching veteran, into this challenging yet rewarding career. My college experience began as a community college student at Lakeland College in Mattoon, Illinois. From there, I received a Bachelor of Science in Mathematics from Eastern Illinois University and a Master of Science in Mathematics from Southern Illinois University. I accepted a tenure-track faculty position at Parkland College, where I have taught developmental and college-level courses for 15 years. I had the opportunity to begin writing textbooks shortly after I started teaching at Parkland. My then department chair and mentor, James W. Hall, and I co-authored several textbooks in Beginning and Intermediate Algebra.

In the fall of 2011, our department began discussing the idea of creating two tracks through our developmental math sequence. The idea stemmed from two issues. First, most of our beginning and intermediate algebra students were headed to either our Liberal Arts Math or our Introduction to Statistics course. Second, we wanted to enhance intermediate algebra to better prepare those students who were headed to college algebra. Obviously, these were two competing ideas! Increasing the algebraic rigor of these courses seemed to "punish" students who were not heading to college algebra. With the two track system, we implemented a solution that best serves both groups of students.

I have to admit that I was initially concerned that offering an alternate path through developmental math for students not planning to take college algebra would lead to a lowering of standards. However, my participation in our committee investigating this idea led me to believe it was possible to offer a rigorous course that was exceedingly more appropriate for this group of students. Since there were no materials for the course, I began creating my own and was paired by McGraw-Hill with Dave Sobecki. Together, we have created the material that I have been using for class testing. After more than a year of piloting these materials and seeing the level of enthusiasm and engagement in the mathematical conversations of my students, I am now convinced that this is an ideal course to refine and offer. As a trusted colleague told me, "this is just a long overdue idea."

Outside of the classroom and away from the computer, I am kept educated, entertained and ever-busy by my wonderful wife Nikki, and our two children, Charlotte, 6 and Jake, 5. I am an avid St. Louis Cardinals fan and enjoy playing recreational softball and golf in the summertime with colleagues and friends.

Dedication: To my kids, Charlotte and Jake.

Acknowledgements

First, we'd like to thank the following individuals who reviewed or class tested *Pathways to Math Literacy*. We thought the original version was pretty good: every person on this list contributed to making it MUCH better.

Amber Anderson, Danville Area Community College
Rachel Anschicks, College of DuPage
Elizabeth Cannis, Pasadena City College
Trey Cox, Chandler-Gilbert Community College
Awilda Delgado, Broward College
Nicole Duvernay, Spokane Community College
Hope Essien, Malcolm X College
Asha Hill, Georgia Highlands College
Linda Hintzman, Pasadena City College
Brandon Huff, Lewis and Clark Community College
Laura Iossi, Broward College
Gizem Karaali, Pomona College
Gayle Krzemien, Pikes Peak Community College
Brian Leonard, Southwestern Michigan College
Christine Mac, Front Range Community College
Catherine Moushon, Elgin Community College
Nicole Munden, Lewis and Clark Community College
Bette Nelson, Alvin College
Peter Nodzenski, Black Hawk College
Karey Pharris, Pikes Peak Community College
Pat Rhodes, Treasure Valley Community College
Doug Roth, Pikes Peak Community College
Jack Rotman, Lansing Community College
Cynthia Schultz, Illinois Valley Community College
Mary Sheppard, Malcolm X College
Craig Slocum, Moraine Valley Community College
Lindsey Small, Pikes Peak Community College
Robin Stutzman, Southwestern Michigan College
Kelly Thannum, Illinois Central College
Ria Thomas, Southwestern Michigan College
Cassonda Thompson, York Technical College
Erin Wilding-Martin, Parkland College

Much of the information we needed to bring this vision to life was provided through a variety of focus groups. So next, we thank our focus group participants, whose valuable insights helped focus our efforts. Guess that's why they call them focus groups.

Patricia Anderson, Arapahoe Community College
Beth Barnett, Columbus State Community College
Ratan Barua, Miami Dade College
Robert Cantin, MassBay Community College
Billye Cheek, Grayson College
Diana Coatney, Clark College
Mahshid Hassani, Hillsbourough Community College
Jessica Lickeri, Columbus State Community College
Faun Maddux, West Valley College
Tanya Madrigal, San Jacinto College
Teri Miller, Clark College
Jeff Morford, Henry Ford Community College
Arumugam Muhundan, State College of Florida
Bill Parker, Greenville Technical College
Betty Peterson, Mercer Community College
Paul Stephen Prueitt, Atlanta Metro State College
Pat Rhodes, Treasure Valley Community College
Wendy Pogoda, Hillsborough CC, South Shore
Saisnath Rickhi, Miami Dade College
Cynthia Roemer, Union County College
Arlene Rogoff, Union County College
Mark Roland, Dutchess Community College
Jorge Sarmiento, County College of Morris
Cathy Schnakenburg, Arapahoe Community College
Pat Suess, St. Louis Community College
Mel Taylor, Ridgewater Community College
Robyn Toman, Anne Arundel Community College
LuAnn Walton, San Juan College
Keith White, Utah Valley University
Valerie Whitmore, Central Wyoming College
Latrica Williams, St. Petersburg College
Mina Yavari, Allan Hancock College

Acknowledgements

Next up is the product team at McGraw-Hill. Without all of these good folks, we'd be just two guys with a good idea and a lot of Word documents.

Ryan Blankenship, Managing Director, who leads the math team through involvement and support.
Holly Rhodes, Brand Manager: If we were a baseball team, Holly would be the manager. Great job, skipper! You share in any success that we have.
Elizabeth O'Brien, Product Developer, who organizes the entire project and graciously fields more emails from us than we could count.
Nicole Lloyd, Director of Digital Content: Nicole's vision and experience as a college math instructor make our digital offerings a huge asset.
Megan Prendergast, Regional Marketing Manager, a new member of the team who spreads the good word to the masses!
Rachel Heiar, Digital Promotions Manager: We have really cool videos and web presence because Rachel is really, really good at what she does...
Jessica Serd, Digital Promotions Intern: ... with Jessica's help.
Peggy Selle, Production Manager, who guides the process that turns a bunch of stuff on our computers into a real book.
Lori Hancock, Content Licensing Specialist: Without Lori's guidance, many of the photos and illustrations that give this book pizazz wouldn't be here.
Tara McDermott, Designer: All of the design elements of this book have their roots in Tara's talent. Thank you!

Finally, above and beyond thanks go to:

Dawn Bercier, who is more responsible than anyone for this project getting off of the ground. Her contributions to the Math team at McGraw-Hill are legendary. Thanks for bringing together peanut butter (Brian) and chocolate (Dave)!

Mary Ellen Rahn, whose leadership guided us through the crucial early stages of an incredibly ambitious project.

Erin Wilding-Martin, who contributed to this product in so many ways that I'm not even sure she can count all of them.

Jack Rotman, a national leader in the Pathways movement, whose thoughtful and in-depth reviews went about eight miles above and beyond.

Geoff Griffiths, the Math department chair at Parkland College, whose leadership and vision made Parkland a pioneer in the Pathways movement.

Nick McFadden, Brian's local McGraw-Hill representative, who listened patiently and provided valuable feedback during the earliest stages of planning for this project.

Jim Hall, who's almost solely responsible for getting Brian into the wild and crazy world of textbook writing.

The students of Parkland College, who contributed an incredible amount of insight by providing feedback on the earliest incarnations of this book.

Cat Sobecki and Nikki Mercer, our wives. We still can't figure out how two math nerds managed to snag these two beauties.

Detailed Table of Contents

Broad Objectives: Assemble the building blocks of functions and equations. Start to develop the idea of function, input/output, independent/dependent. Practice more number sense (including probability and percent chance). Carefully explore the meaning of a variable and develop the idea of a solution of an equation, solve some basic equations and inequalities, and develop a problem solving strategy.

Unit 4: Living in a Nonlinear World 231

Broad Objectives: Expand to nonlinear relationships. Study normally distributed data, the Pythagorean theorem and distance formula, applications based on graphs of quadratic and exponential functions, quadratic and exponential regression, inverse variation, scientific notation, operations with polynomials, factoring (stressing the relationship between factors, x-intercepts, and solutions), and algebraic approaches to solving quadratic equations.

Unit 1
Numbers and Patterns

Outline

Lesson 1-1 Where Does the Time Go?

Learning Objectives

- ☐ 1. Complete and analyze a weekly time chart.
- ☐ 2. Compute percentages.
- ☐ 3. Create and interpret pie charts using a spreadsheet and by hand.

Time is what we want most, but what we use worst.

– William Penn

One of the most important aspects of success in college is very underrated: learning how to manage time. This is a skill that you're unlikely to acquire by chance: in order to understand how to use your time most effectively, you have to first become aware of how you're using your time. Then you'll have to develop a plan that will help you best take advantage of your valuable time. This can help you to do better in your classes, and can also help you to have more time to do the things you enjoy.

0. After reading the opening paragraph, what do you think the main topic of this section will be?

1-1 Class

Here's a sample time chart put together by a student who was interested in identifying the amount of time she spent for one week. Each hour is marked with C (time in class), H (time spent on homework), S (sleep), W (work) or O (other time commitments). In the Portfolio portion of this lesson, you'll be asked to fill out a similar time chart of your own. For now, we'll analyze this chart.

	Sunday	Monday	Tuesday	Wednesday	Thursday	Friday	Saturday
12am-1:00	O	S	S	S	S	S	O
1:00-2:00	O	S	S	S	S	S	O
2:00-3:00	S	S	S	S	S	S	S
3:00-4:00	S	S	S	S	S	S	S
4:00-5:00	S	S	S	S	S	S	S
5:00-6:00	S	S	S	S	S	S	S
6:00-7:00	S	S	O	S	O	S	S
7:00-8:00	S	S	W	S	W	S	S
8:00-9:00	S	O	W	O	W	O	S
9:00-10:00	O	C	W	C	W	C	S
10:00-11:00	O	C	W	C	W	C	S
11:00-Noon	H	H	H	H	H	H	H
Noon-1:00	H	O	C	O	C	O	H
1:00-2:00	W	C	C	C	C	C	O
2:00-3:00	W	H	C	H	C	H	O
3:00-4:00	W	O	H	O	H	O	O
4:00-5:00	W	O	H	O	H	O	H
5:00-6:00	W	O	O	O	O	O	H
6:00-7:00	W	O	O	O	O	O	H
7:00-8:00	W	H	H	H	H	O	O
8:00-9:00	H	H	H	H	H	O	O
9:00-10:00	H	O	H	O	H	O	O
10:00-11:00	H	O	O	O	O	O	O
11:00-12am	O	S	O	S	O	O	O

1. Count the number of spaces containing each letter in the time chart (C, H, S, W, O). Use the results to fill in the following chart, writing the total number of hours devoted to each activity during one week.

	Hours
Class	15
Homework	32
Sleep	54
Work	15
Other	52

2. Without adding the hours in the table, how can you decide what the total number of hours should be?

24 HOURS PER DAY
7 DAYS PER WEEK

$24 \times 7 = 168$

3. Most college advisors will tell you that a good rule of thumb is to allow 2 hours of study time outside of class for each hour spent in class. Is the student that filled out the table in Question 1 following that advice?

15 HRS CLASS WORK
30 HRS STUDY TIME

$15 \times 2 = 30$
ANSWER IS YES

4. Use your results from Questions 1 and 2 to fill in the next chart with the percentage of hours in a week devoted to each activity. (Round to the nearest percent.)

	Percentage
Class	9 %
Homework	19 %
Sleep	32 %
Work	9 %
Other	31 %

$15/168 = .0892 = 8.9 = 9\%$
$32/168 = .1904 = 19\%$
$54/168 = .3214 = 32.1 = 32\%$
$52/168 = .3095 = 31\%$

Math Note

Recall that to find a percentage of a total, you divide the given number by the total, then multiply the result by 100.

5. Studies have shown that college students get about 6 ½ hours of sleep per night on average, but that academic achievement improves when students average 8 hours of sleep. What percentage of the hours in a week would be devoted to sleep if you get 6 ½ hours of sleep per night? What if you get 8 hours? How is our friend from Question 1 doing when it comes to sleep?

$6.5/24 = 0.2708 = 27\%$

$8/24 = 0.333 = 33\%$

HE IS DOING O.K.

6. A circle can be divided into 360 equal units of measure, which we call **degrees**. In other words, an entire circle is made up of 360° (the ° symbol represents degrees). Multiply the decimal form of each percentage in Question 4 by 360°, and put the results into the next chart. Round to the nearest whole degree.

	Degrees
Class	32°
Homework	69°
Sleep	116°
Work	32°
Other	111°

A **pie chart** is a diagram used to compare the relative sizes of different parts of a whole. For example, the pie chart to the right, adapted from a USA Today survey, describes people's hand-washing tendencies when using a public restroom. In this case, the whole is the number of people surveyed, and each of the parts is the percentage of folks that gave a certain response. Since 70% of respondents said "Always," the "Always" category fills 70% of the circle. This represents $0.7 \times 360° = 252°$. (See the next page for a quick review of percents.) The "Sometimes" category fills 29%, which corresponds to $0.29 \times 360° \approx 104°$, and the "Never" category just 1%, corresponding to $0.01 \times 360° \approx 4°$.

How Often Do You Wash Your Hands After Using a Public Restroom?

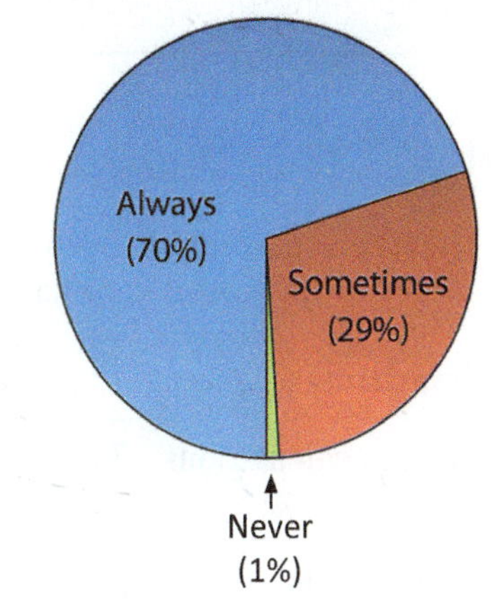

Now let's see if we can build a pie chart that illustrates the amount of time spent on various activities for the student survey that began this section.

7. Build a pie chart for the information in Question 6 by marking off angles on the circle to the right, starting at the 0° mark, that correspond to the numbers of degrees in Question 6. In this case, the "whole" refers to the total number of hours in a week, and each part is the portion of that time spent on one of the activities. Each of the light gray lines on the graph represents 10°.

8. Does the pie chart make it easy to analyze the amount of time this student spends on each activity? Explain.

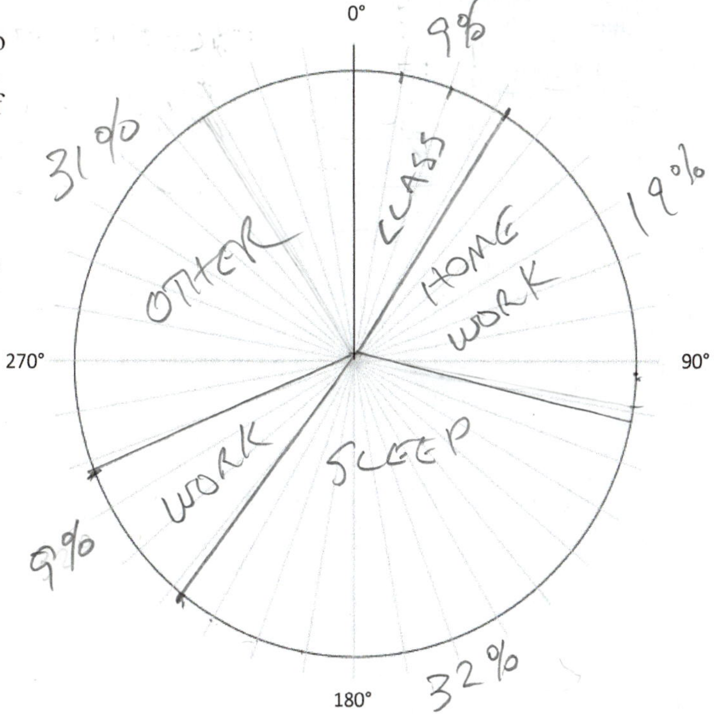

In preparation for the group portion of this section, everyone in your class should supply their answer to the one-question poll below. The results will be tabulated and analyzed in the group work.

How do you feel about math in general?	
I love it	☑
I like it	☐
I can take it or leave it	☐
I don't like it	☐
I hate it	☐

Review of Percents

1. The word "percent" literally means "per hundred." When we read that 70 percent of respondents to a survey always wash their hands in a public restroom, it means that 70 people per hundred do so.

2. Translating a percent as "per hundred" makes it easy to convert percents to fraction form for using them in calculations: 70% means 70 per hundred, or $\frac{70}{100}$. And there's your fractional form for 70%!

3. This also shows us how to convert percents to decimal form: dividing any number by 100 moves the decimal two places to the left. That makes 70% equal to 0.70, 5% equal to 0.05, and 23.5% equal to 0.235.

4. If you forget to convert a percent into fractional or decimal form when doing a calculation, you can usually catch your mistake with a bit of thought. In the pie chart on the previous page, if we used $70 \times 360°$ instead of $0.7 \times 360°$, we'd have found an angle of 25,200°, which is pretty silly given that a full circle is 360°.

1-1 Group

1. Numerous studies have shown that one of the best ways to do better in college classes is to study in pairs or groups. To help you get started, if you feel comfortable sharing contact information, exchange the information in the table below. The group you're in now will be your small group for the first unit of this course. When you get used to meeting in class, you'll likely find that meeting outside of class to study and work on homework is a good idea as well, so include some study times that would be convenient for you to meet.

Name	Phone Number	Email	Available times
SHAW	971-331-3440	SHAWN_MARTIN@YAHOO	
SALLY	503-621-7216	DAVESALLY 503 @ GMAIL	
CHEYENNE	802-349-6407	CHEYENNE. PIERCE	@ PCC.EDU
JOE	503-466-9827	JOSEPH.MABBOTT 1 @	PCC.EDU
SARAH	971 283 3207	sarah.hohstadt@ pcc. edu	

VWSarah22@gmail.com

2. Use data from the class poll on page 6 about what students think about math to complete the table. Then use your results to draw a pie chart comparing the responses.

Feelings toward math	Number	Percent	Degrees
Love it	2	13%	47°
Like it	0	0%	0°
Take or leave	8	53%	192°
Don't like	2	13%	47°
Hate it	3	20%	72°

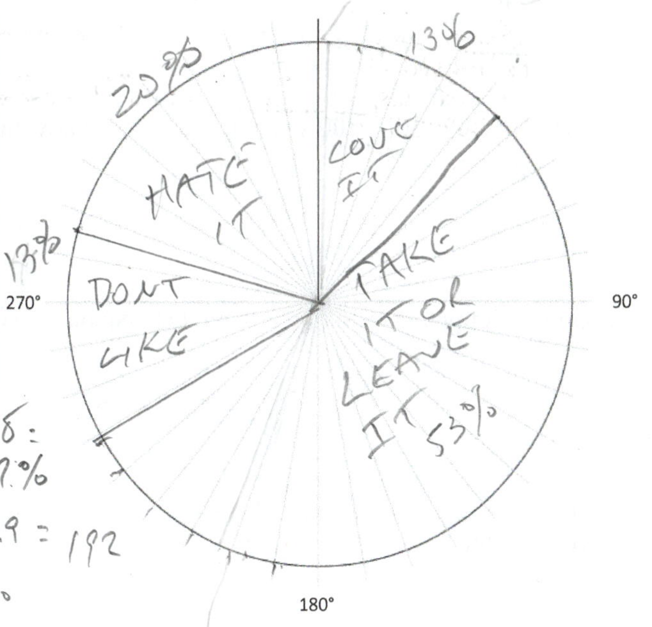

2/15 = .13333 = 13% .13333 × 360 = 46.8 = 47%

8/15 = .53333 = 53% .5333 × 360 = 191.9 = 192

3/15 = .2 = 20% .2 × 360 = 72°

Using Technology: Creating a Pie Chart

To create a pie chart in Excel:

1. Type the category names in one column (or row).
2. Type the category values in the next column (or row).
3. Use the mouse to drag and select all the data in those two columns (or rows).
4. With the appropriate cells selected, click the **Insert** tab, then choose **Chart** and Pie chart. There are a few different styles you can experiment with, but starting with the simplest is a good idea.

You can add titles, change colors and other formatting elements by right-clicking on certain elements, or using the options on the **Charts** menu. Try some options and see what you can learn!

See the Lesson 1-1 video in class resources for further instruction.

Interpreting Pie Charts

A national survey on college students and work was conducted during the 2003–2004 academic year and reported on the website www.acenet.edu. The results are summarized in the pie chart.

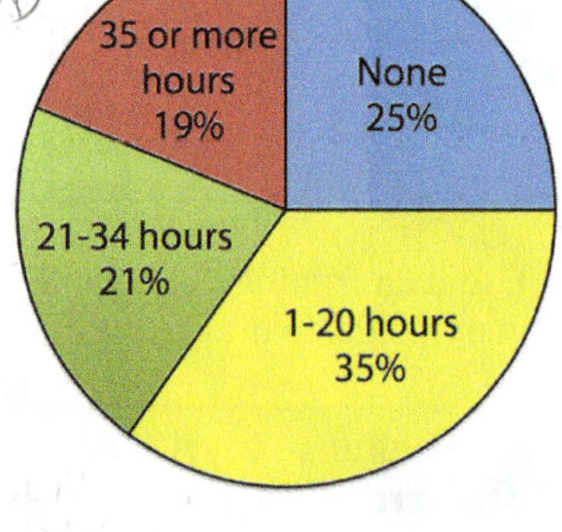

3. What range of hours worked was most common?

1-20 HRS WAS THE MOST COMMON HOURS WORKED

4. What percentage of students worked between 21 and 34 hours per week?

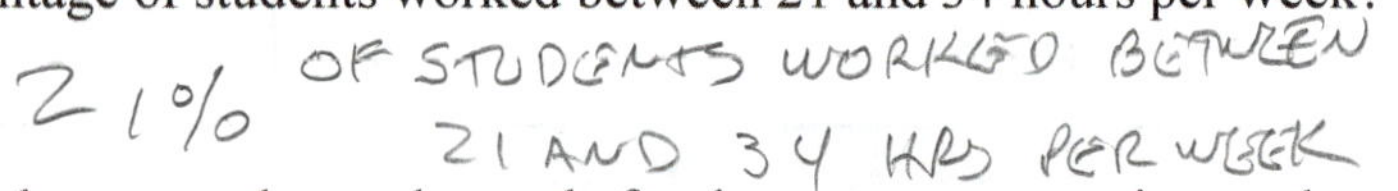

5. How many degrees make up the angle for the sector representing students who worked from 1 to 20 hours per week?

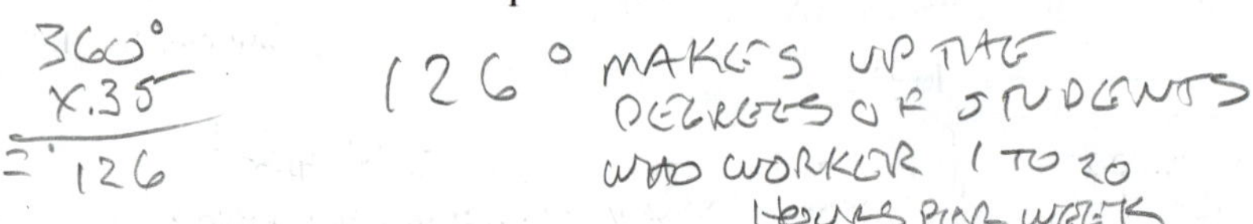

6. On a campus with 7,500 students, on average how many would you expect to work at least 1 hour per week?

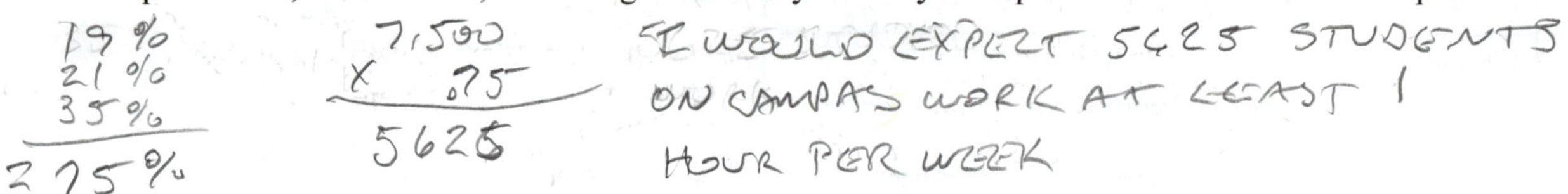

7. On the campus in Question 6, how many students would you expect to work less than 21 hours?

35% 7500
+25% x.60
=60 = 4500

I WOULD EXPECT 4500 STUDENTS WORK LESS THAN 21 HOURS PER WEEK

8. How do you think these results would compare with a similar survey conducted for this school year? Why do you feel that way?

I THINK THE NUMBERS OF HOURS STUDENTS WORKED PER WEEK WOULD INCREASE.

I FEEL THAT STUDENTS WOULD WORK MORE DURING THE SUMMER TERM

1-1 Portfolio

Name ______________________________

Check each box when you've completed the task. Remember that your instructor will want you to turn in the portfolio pages you create.

Technology

1. ☐ Fill in a time chart like the one at the beginning of this lesson based on your schedule. Then compile the results and use Excel to create a pie chart showing where your time is spent. There's a sample template in the online resources for this lesson to help you get started. Copy and paste the pie chart into a Word document, and write a couple of sentences below the chart describing any thoughts you have about what you learned.

Skills

1. ☐ Include any written work from the online skills assignment along with any notes or questions about this lesson's content.

Applications

1. ☐ Complete the applications problems.

Reflections

Type a short answer to each question.

1. ☐ Do you spend your time efficiently? Explain.
2. ☐ Why are you in college? If you've never thought about a good answer to this question, you certainly should have!
2. ☐ What changes do you plan to make in how you budget your time?
3. ☐ What resources do you plan to use to help you do well in this course?
4. ☐ List some benefits of working/studying in groups rather than alone.
5. ☐ Take another look at your answer to Question 0 at the beginning of this lesson. Would you change your answer now that you've completed the lesson? How would you summarize the topic of the lesson now?
6. ☐ What questions do you have about this lesson?

Looking Ahead

1. ☐ In preparation for the next lesson, visit the learning styles website linked under Lesson 1-2 resources, and take the learning style assessment. Your results will be recorded in a table in the Class portion of the next lesson.
2. ☐ Read the opening paragraph in Lesson 1-2 carefully, then answer Question 0 in preparation for that lesson.

1-1 Applications Name ____________________

Read the article on pizza franchises from the Lesson 1-1 online resources. Answer each question based on the article.

1. How many slices of pizza are eaten per second in the United States? What about per hour? (Hint: There are 3,600 seconds in an hour.)

350 PER SECOND
X 3600 IN AN HOUR = THERE ARE 1260000 PER HOUR

2. How many independent pizza stores are there in the United States? Show work to justify your answer.

67,554 STORES X 59% =
X .59
= 39,856.86 = THERE ARE 39857 IN THE US

3. How many degrees make up the angle for the sector representing sales by other top chains? How many degrees make up the angle for the number of stores for other top chains?

SALES 360°
X 15%
64.8

STORES 360°
X .15
54°

4. According to the article, Pizza Hut has over 7,500 stores in the United States. Based on the pie chart representing number of stores, how many stores does Pizza Hut have? Show work to justify your answer. Compare your result to the number given in the article, and discuss possible reasons for discrepancies.

SALES 360%
X .15

54° IN SECTOR FOR STORES

5. For every $20 spent on average on pizza in the United States, how much goes to chain pizzerias?

FOR EVERY $20 SPENT $9.80 GOES TO INDEPENDANT STORES

6. Compare the percentage of independent pizza stores to the percentage of sales. What do you think the difference indicates?

THE PERCENTAGE OF SALES IS LOWER COMPARED TO THE PERCENTAGE OF INDEPENDENT STORES

THE CHAIN STORES ARE MAKING A HIGHER PERCENTAGE OF SALES

Lesson 1-2 It's All About Style

Learning Objectives

- ☐ 1. Identify and understand your learning style.
- ☐ 2. Create and interpret bar graphs using a spreadsheet and by hand.

Make your mistakes work for you by learning from them.

– Donald Trump

Yes friends, this lesson is about style – learning style, that is. Learning is an active process, not a spectator sport. Think of some other active processes in your life: do you walk the same as everyone else? What about talk, sing, dance, drive, eat, smile, laugh... well, you get the picture. The answer, of course, is no. Then why on earth would you think that you learn like everyone else? Part of getting an education is learning how you learn. In this lesson, we'll study some of the different ways that people learn.

0. After reading the opening paragraph, what do you think the main topic of this section will be?

1-2 Class

Every student can benefit from finding out a little bit about his or her learning style – in almost any area, understanding your strengths and weaknesses is an important part of improving.

1. Record the scores you found when taking the online learning style assessment from the online resources for Lesson 1-2.

Style	Score	Style	Score
Visual (spatial)	5	Verbal (linguistic)	5
Social (interpersonal)	6.2	Solitary (intrapersonal)	7.6
Physical (kineshthetic)	1.6	Logical (mathematical)	7.6
Aural (auditory-musical)	5		

A **bar graph** is a visual way to compare the sizes of different values. Pie charts are great for comparing parts to a whole: bar graphs are effective for comparing parts to other parts. In a bar graph, the length of each bar gives an indication of how the size of a quantity compares to other related quantities. The bar graph to the left shows the percentage of global Internet users that visit the top five most-visited sites in the world. Just by looking at the heights of the bars, we can see how the percentages for these sites compare. By looking more closely at the scale on the vertical axis, we can get a good estimate of the actual percentage for each of the top five sites.

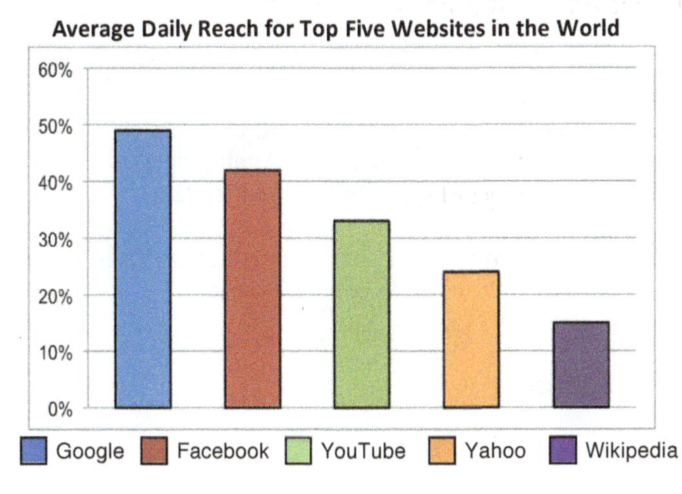

2. Which site has the highest daily reach? What percentage of Internet users go that site on an average day?

3. Use your learning style scores from Question 1 to draw a bar graph by creating a rectangle for each style with height corresponding to your score for that style.

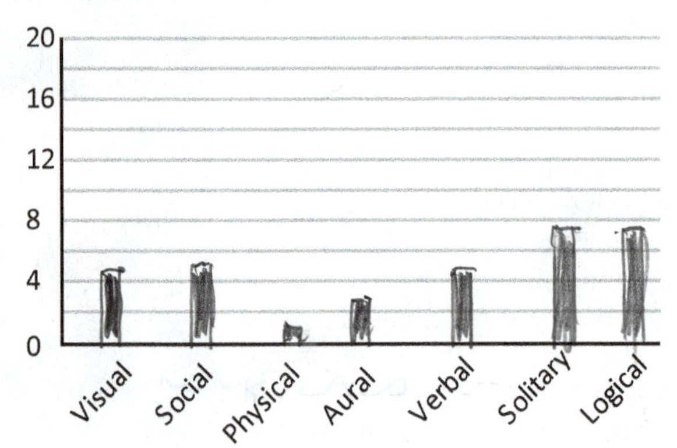

4. Is there one style that stands out as having a higher score than the others for you? How does its bar compare in size to the next highest? NO FAIRLY EVEN OVER 5/7th

5. Describe how the bar graph makes it easier to compare the relative scores than the table from Question 1 did.

Using Technology: Creating a Bar Graph

To create a bar graph in Excel:

1. Type the category names in one column (or row).
2. Type the category values in the next column (or row).
3. Use the mouse to drag and select all the data in those two columns (or rows).
4. With the appropriate cells selected, click the **Insert** tab, then choose **Chart** and Column or Bar graph. ("Column" gives you vertical bars, "Bar" gives you horizontal.) Again, there are a few different styles you can experiment with, but starting with the simplest is a good idea.

You can add titles, change colors and other formatting elements by right-clicking on certain elements, or using the options on the **Charts** menu. Try some options and see what you can learn!

See the Lesson 1-2 video in class resources for further instruction.

1-2 Group

The Harris Interactive polling agency surveyed 2,094 consumers in October of 2012, asking which former president they would most enjoy having a burger with. The top four vote-getters are summarized in the bar graph.

1. Which former president had the highest percentage, and what was the percentage?

CLINTON – 22%

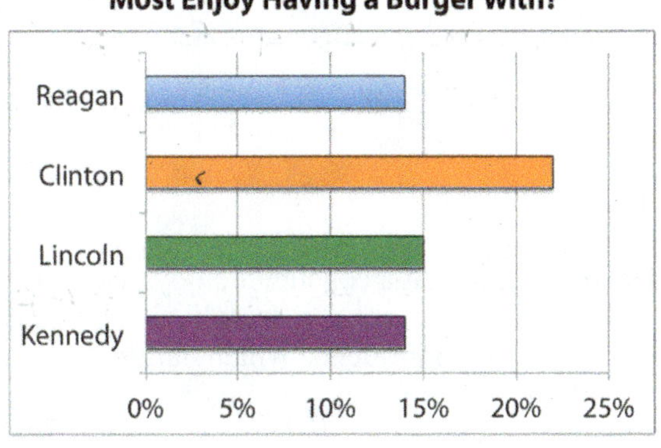

2. How many of the respondents would most enjoy having a burger with the most popular choice?

2094 X 22% = 2094 X .22 = 460.68

= 461 PEOPLE

3. Which two presidents tied in the poll?

REAGAN & KENNEDY

4. How many people chose a president not in the top four?

14% + 14% + 15% + 22% = 65% =

.14 + .14 + .15 + .22 = .65

.35 X 2094 = 732.9 = 733

The yearly net income or loss for the Ford Motor Company from 2007 to 2011 is shown in the next graph.

Ford Net Profit/Loss

Billions of Dollars: 2007: −3; 2008: −15; 2009: 3; 2010: 7; 2011: 20

5. Describe how Ford performed in 2008 and in 2010 in terms of profit or loss.

2008 = −15 BILLION DOLLARS

2010 = +7 BILLION DOLLARS

6. What were the best and worst years for Ford from 2007 to 2011? Explain how you decided.

2010 = 20 BILLION DOLLAR INCREASE

2008 = 15 BILLION " DECREASE

BY SEEING THE LEAST AND MOST GAINS

7. What was the largest single-year increase in profit?

2009 = SHOWS A 18 BILLION DOLLAR INCREAS

−15 → +3 = 18

8. A mathematical way of expressing profit is to use a ___+___ number.

A ___—___ number is used to express a loss.

Math Note

Sometimes a bar graph will include the actual value at the end of the bars, making it easier to read accurately.

It's really important to pay attention to all aspects of graphs, not just what the pretty picture looks like. Sometimes graphs can be misleading, as seen in the next example. These graphs might be poorly drawn, or they may be intentionally misleading to try to influence your opinions, or get you to buy products that you otherwise wouldn't.

The first bar graph below compares the number of calories in a regular cheeseburger at the three top burger franchises in the United States.

Calories in a Regular Cheeseburger at the Three Top Burger Chains

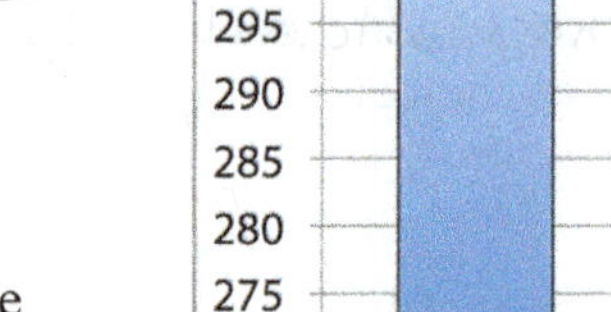

9. Look only at the HEIGHTS of the bars on the first graph. Describe how the number of calories appears to compare based on those heights.

IT APEARS THAT # 1 HAS ALOT MORE CALORIES IN IT

10. Assuming that a consumer enjoys the taste of all three cheeseburgers equally, how likely do you think this graph would be to influence which burger they choose to eat most often? I BELIEVE THAT MOST PEOPLE WOULD CHOOSE # 1

Calories in a Regular Cheeseburger at the Three Top Burger Chains

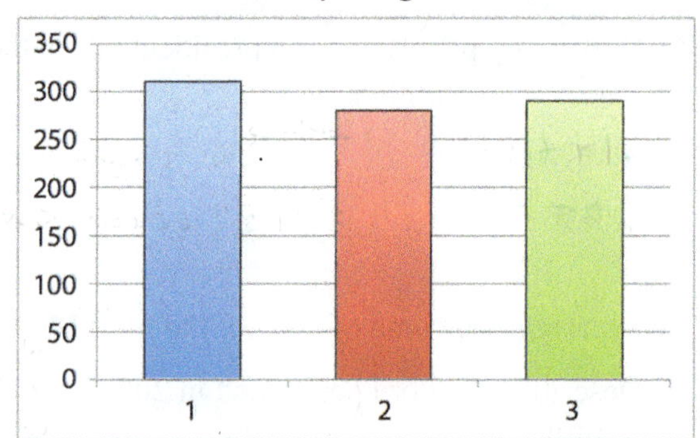

11. Does the second graph provide the same information as the first? NO, THE SECOND GRAPH SEEMS TO BE MORE REALISTIC

12. Again focusing on the heights of the bars, how do the three burgers compare in terms of calorie content?

ALL 3 SEEM TO BE FAIRLY CLOSE

13. What scale is used on the vertical axis for each graph? (**Scale** refers to the spacing of the labels on an axis.)

SCALE #1 SHOWS A HIGHLITED GRAPH FROM 270-310 WHILE #2 SHOWS A OVERALL GRAPH FROM 0-350

14. Which of the three franchises would be most likely to use the first graph in advertising? Which would be most likely to use the second? Needless to say, you should explain your answers.

#1 AS IT SHOWS A MORE DRAMATIC

1-2 Portfolio

Name WILLIAM BECKER

Check each box when you've completed the task. Remember that your instructor will want you to turn in the portfolio pages you create.

Technology

1. ☐ Find an example of a bar graph in a book, magazine, newspaper, or online article on a subject that you find interesting. Recreate the bar graph in a spreadsheet, including labeling. Then copy and paste your bar graph into a Word document and type a sentence or two explaining why you find the topic interesting, and what the bar graph tells you. Be sure to make note of your source!

Skills

1. ☐ Include any written work from the online skills assignment along with any notes or questions about this lesson's content.

Applications

1. ☐ Complete the applications problems.

Reflections

Type a short answer to each question.

1. ☐ Write a sentence or two about each of the seven learning styles. Describe each in your own words.
2. ☐ Were you surprised by what your primary learning style is? Why or why not?
3. ☐ How do you plan to take advantage of knowing your primary learning style?
4. ☐ Take another look at your answer to Question 0 at the beginning of this lesson. Would you change your answer now that you've completed the lesson? How would you summarize the topic of this lesson now?
5. ☐ What questions do you have about this lesson?

Looking Ahead

1. ☐ To prepare for the next lesson, you'll need to take a personality test. (You do have a personality, right?) Links to several such tests can be found under Lesson 1-3 resources. Your personality type will be a four-letter code. Make note of it.
2. ☐ Read the opening paragraph in Lesson 1-3 carefully, then answer Question 0 in preparation for that lesson.

1-2 Applications

Name ______________________

In Questions 1 and 2, two different situations are described that you could illustrate with a graph. For one of the two, it would be most appropriate to use a bar graph, and for the other a pie chart would be a better choice.

Social Network Use: A survey of 100 college freshman found that 82 use Facebook, 10 use MySpace, 48 use Twitter, 60 use Instagram, and 33 use Google+.

Test scores: In a Geology class with 32 students, 6 earned an A on the first exam; 12 earned a B; 7 a C; and 4 earned a D.

1. Which data is best illustrated with a bar graph? Explain why you feel that way, then draw a bar graph representing the data.

THE SOCIAL NETWORK USE, BECAUSE MUTIPLE STUDENT USED MULTIPLE NETWORKS MAKING UP OVER 100% WICH WOULD BE NONSENSE IN A PIE CHART

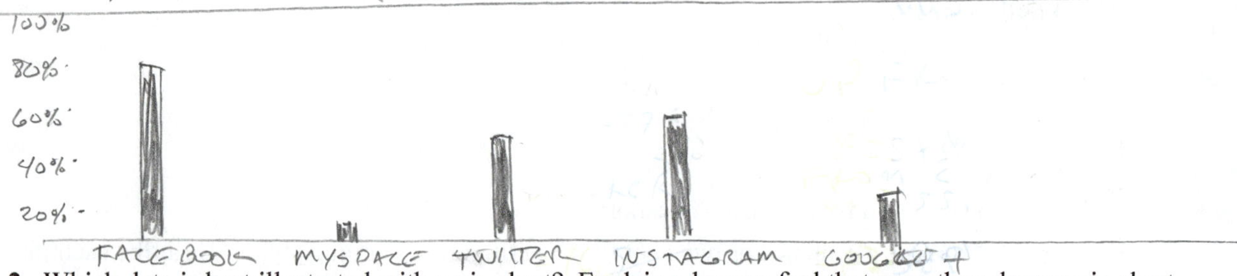

2. Which data is best illustrated with a pie chart? Explain why you feel that way, then draw a pie chart representing the data.

TEST SCORES, BECAUSE THE NUMBER OF ITEM'S AND VARIATIONS DO NOT EXCEED 100%

$\frac{6}{32} = .1875 = 18.75\% = 19\%$

$\frac{12}{32} = .375 = 37.5\% = 38\%$

$\frac{7}{32} = .21875 = 21.875\% = 22\%$

$\frac{4}{32} = .125 = 12.5\% = 13\%$

$.19 \times 360 = 68.4 = 68°$

$.38 \times 360 = 139.84 = 140°$

$.22 \times 360 = 79.2 = 79°$

$.13 \times 360 = 46.8 = 47°$

334?

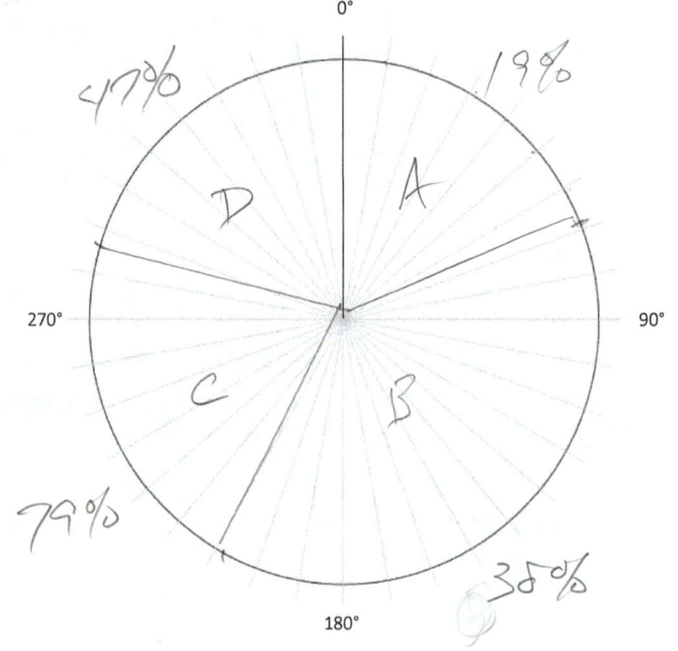

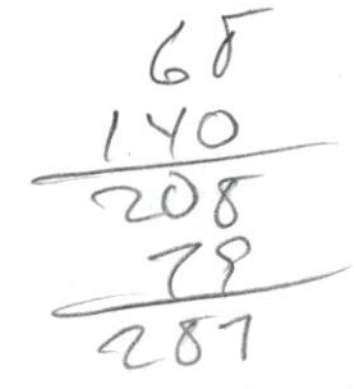

1-2 Applications Name WILLIAM BEEZER

Sometimes bar graphs will have more than one bar in each category. This is useful when categories can be divided into subcategories. An example of this is shown in the graph below, which was reproduced from the website of the Substance Abuse and Mental Health Services Administration, a division of the U.S. Department of Health and Human Services. A study was done on alcohol use among people in the 12–20 age bracket in 2010, with some results summarized by the graph. The SAMHSA website defines alcohol use as follows:

Current (past month) use - At least one drink in the past 30 days.

Binge use - Five or more drinks on the same occasion (i.e., at the same time or within a couple of hours of each other) on at least 1 day in the past 30 days.

Heavy use - Five or more drinks on the same occasion on each of 5 or more days in the past 30 days.

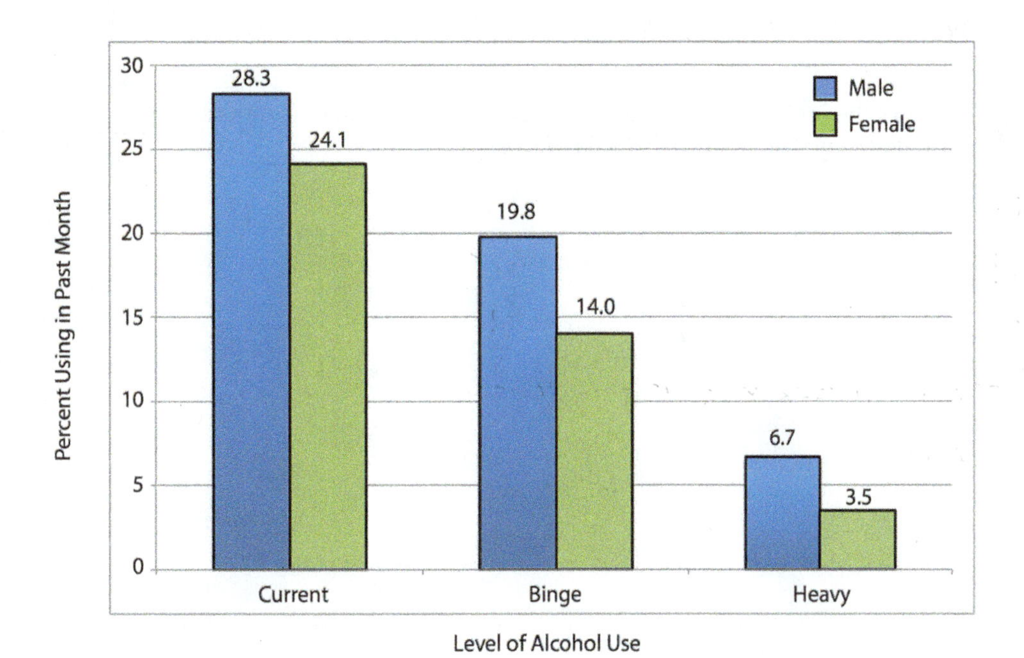

3. What does the graph indicate about the difference in alcohol usage between boys and girls?

THAT BOYS HAVE A SLIGHTLY HIGH % OF USE ACROSS ALL CATAGORIES

28.3 - 24.1 = 4.2
19.8 - 14.0 = 5.8
6.7 - 3.5 = 3.2 / 13.2 X .333 = 4.3956 = 4.4 % DIFFERANCE

4. Which of the drinking categories shows the greatest discrepancy between boys and girls?

BINGE IT HAS HIGHEST % OF DIFFERNCE

28.3 - 24.1 = 4.3%
19.8 - 14.0 = 5.8%
6.7 - 3.5 = 3.2%

5. What percentage of boys reporting current alcohol use were in the heavy drinking category? What about girls?

6. Based on percentages, if a person aged 12–20 is a current drinker, are they more likely to be a binge drinker if they are a boy or a girl? Explain how you got your answer.

7. The next graph shows the results of a similar survey conducted in 2007. Compare the two graphs and write at least two trends in underage alcohol use that you notice.

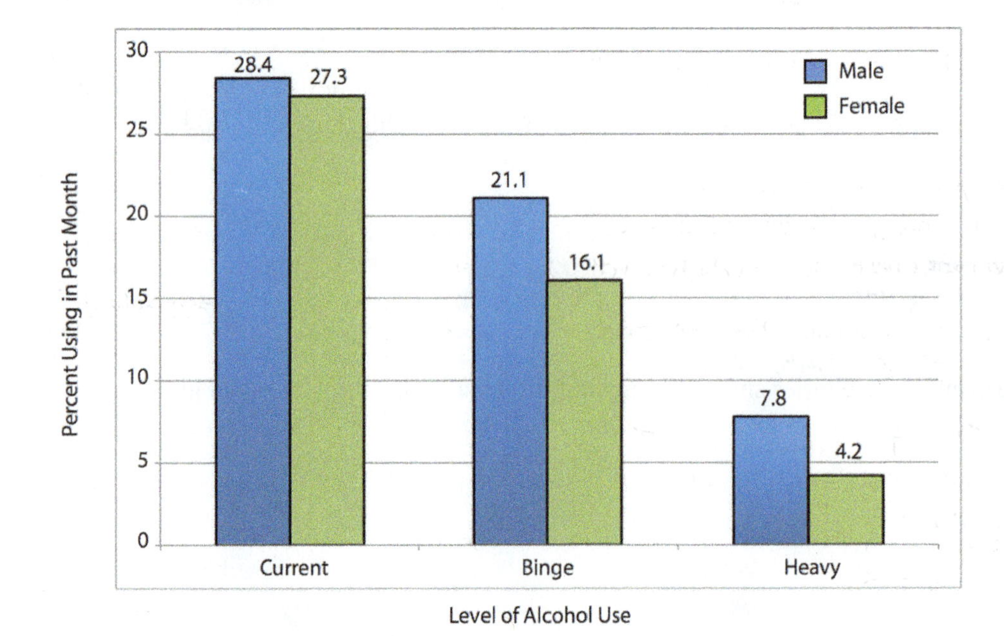

Lesson 1-3 What's Your Type?

Learning Objectives

- ☐ 1. Consider the role your personality type plays in how you interact with others and how you learn.
- ☐ 2. Create and interpret Venn diagrams.

The meeting of two personalities is like the contact of two chemical substances; if there is any reaction, both are transformed.
– Carl Gustav Jung

In almost every walk of life, from social, to academic, to professional, working with and interacting with other people is important. But how many people actually take the time to think and learn about what may be the single most important aspect of their lives? In my view, one of the best things about higher education is being exposed to people with different backgrounds and ideas. When you learn to embrace the value of diversity, a new world of learning and social opportunities opens before you. In this lesson, we'll learn about different personality types and think about how your type affects the way you interact with others in a learning environment.

0. After reading the opening paragraph, what do you think the main topic of this section will be?

1-3 Class

In preparation for this class, you should have taken an online personality assessment. The result is a four-letter code that describes your specific personality type. The code is made up of a combination of 8 letters in four pairs, so we'll begin by learning what each letter represents.

First pair: E (Extrovert) or I (Introvert)

Extroverts draw energy from action, preferring to act, then reflect, then adjust actions based on that reflection. Introverts expend energy from action, and prefer to think first, then act upon the thoughtful planning that they've done.

Second pair: S (Sensing) or N (Intuition)

Those with a preference for sensing prefer to base their thoughts and actions on concrete things that they can observe with their five senses; they distrust hunches and "gut feelings." The intuition people tend to trust information that is more abstract or theoretical, and information gathered from previous thoughts and experiences shared by others.

Third pair: T (Thinking) or F (Feeling)

Thinking and feeling are known as the decision-making functions; both are used to make conscious, rational decisions based on the data received from either sensing or intuition. Those with a thinking preference tend to make decisions based on what seems most logical and reasonable with a bit less regard for emotional impact. The feeling preference indicates a tendency toward decision-making from a more emotional and empathetic viewpoint, trying to achieve balance and consensus among all people affected by a decision.

Fourth pair: J (Judging) or P (Perceiving)

These types are about the style you prefer when interacting with the outside world. A person with the judging preference is most likely to approach actions with a plan in place, preferring organization, preparation, and staying on schedule. The perceiving style folks are perfectly fine with "playing it by ear," making up the plan as they go. They're flexible and adaptive, and like new opportunities and challenges to pop up.

Now that we know a little bit about personality characteristics, we'll use a new type of graph known as a **Venn diagram** to see how your personality type compares to others, and later you'll reflect on how this might affect working with others in groups. Venn diagrams are used to represent sets of objects (in this case the letters from the Myers Briggs personality test) and visualize what they have in common, and how they differ.

1. Write the four-letter code for your personality type here: ___________

2. Pair up with someone else in the class, and write your partner's code here: ___________
   (It would be a good idea to pick someone that doesn't have the exact same type as you.)

3. Label the two circles in the Venn diagram with one of your names on each line.

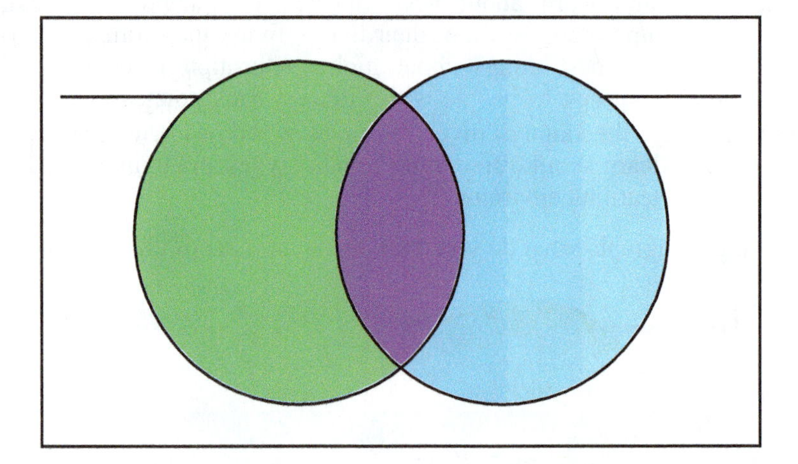

4. Write the letters your personality type has in common with your partner's in the overlapping portion of the two circles.

5. Write any letters in your personality type but NOT your partner's inside the circle labeled with your name, but outside the overlapping portion.

6. Write any letters in your partner's personality type but NOT yours inside the circle labeled with your partner's name, but outside the overlapping portion.

7. Write the remaining letters that don't appear in either code outside the circles, but inside the rectangle.

8. This Venn diagram shows which personality traits you have in common with your partner, and which you each have separately. Write a description of what the diagram says to you.

Math Note

The objects inside a circle in a Venn diagram represent all elements of the set represented by that circle. The objects *outside* make up the **complement** of that set.

1-3 Group

We can also draw Venn diagrams with three circles to study interactions between three different sets. In this activity, you'll compare characteristics with two other people in your group.

1. Complete the following Venn diagram by placing the letter next to each statement in the appropriate location. For example, if the two people corresponding to the top two circles like to study late at night and the third doesn't, you should put A in the portion marked with the arrow.

A. Like to study late at night
B. Like to study early in the morning
C. Like to get work done early
D. Like to put work off to the last minute
E. Will miss very few classes
F. Usually miss a lot of classes
G. Working a job for 20 or more hours per week
H. Working a job for less than 20 hours per week
I. Taking 15 hours or more of classes
J. Taking less than 15 hours of classes
K. Have children
L. Do not have children
M. Feel confident in most math courses
N. Do not feel confident in most math courses

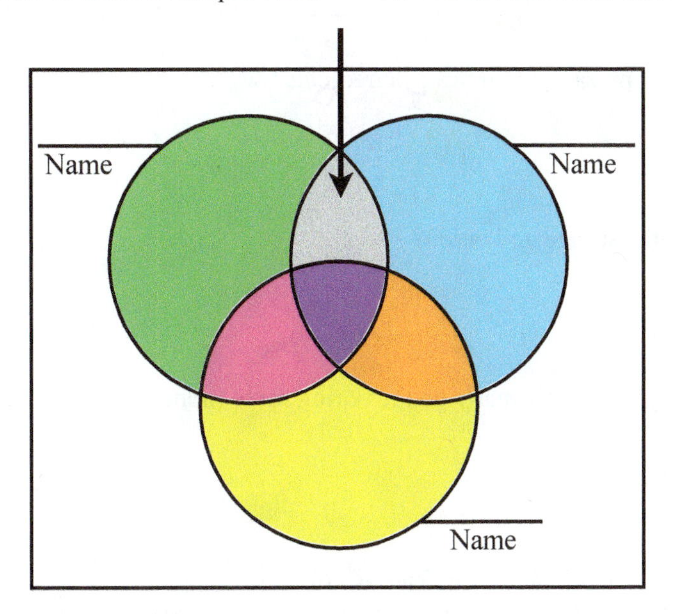

2. Write any observations you find interesting from the completed diagram.

Math Note

It's possible to draw Venn diagrams with more than 3 circles, but it gets pretty complicated, making them kind of difficult to interpret.

According to a survey conducted by the National Pizza Foundation that I just now made up, out of 113 customers surveyed, 47 prefer pizza with pepperoni, 56 prefer sausage, and 30 prefer onion. These results are displayed in the next Venn diagram.

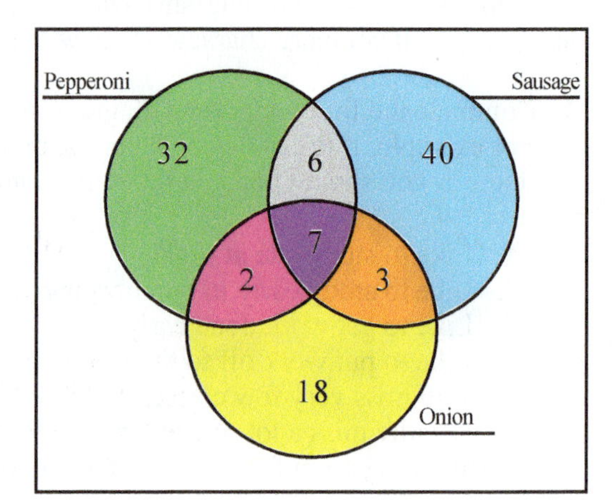

3. According to the information given in the previous paragraph, 47 customers prefer pizza with pepperoni. But notice that 47 doesn't appear in the diagram anywhere. Why not?

4. How many customers prefer pizza with only sausage (and none of the other two ingredients)?

5. How many customers prefer pizza with all three ingredients?

6. How many prefer pizza with sausage and onion?

7. I hate to call anyone boring, but... how many go the boring route (none of these items)?

In a study of 400 entrees served at 75 campus cafeterias, 70 had less than 12 grams of fat but not less than 350 calories; 48 had less than 350 calories but not less than 12 grams of fat; 140 had over 350 calories and over 12 grams of fat. Fill in the Venn diagram with the given information, then use it to answer the next three questions.

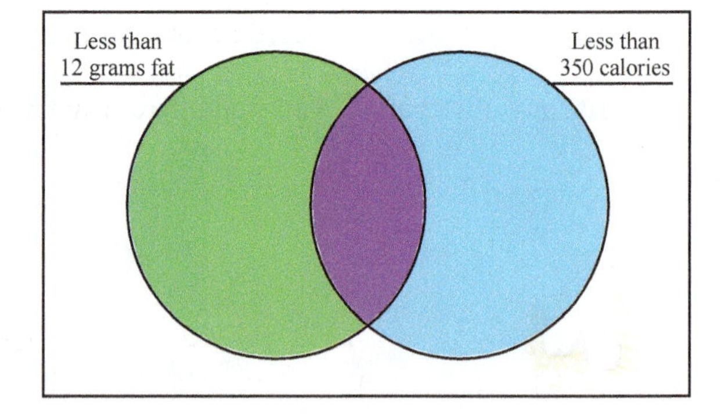

8. How many meals had less than 12 grams of fat and less than 350 calories?

9. How many meals had less than 350 calories?

10. How many had less than 12 grams of fat?

1-3 Portfolio Name ____________________

Check each box when you've completed the task. Remember that your instructor will want you to turn in the portfolio pages you create.

Technology

1. ☐ A real Myers Briggs personality test would be administered by a qualified professional; the free online versions are based on the real test. Do a Web search for information on your specific personality type. Do you agree or disagree with the result of your test? In what ways?
2. ☐ Find a Venn diagram online based on a topic that interests you. Copy and paste the image into your portfolio, then explain why you found it interesting, and what you learned from it.

Skills

1. ☐ Include any written work from the online skills assignment along with any notes or questions about this lesson's content.

Applications

1. ☐ Complete the applications problems.

Reflections

Type a short answer to each question.

1. ☐ Think about what you learned about yourself from studying your personality type. How can you use this information in your academic and social lives?
2. ☐ Now consider your personality type as well as the Venn diagram on page 21. How can you use what you learned to help you in this class, and other classes where group work is important?
3. ☐ Take another look at your answer to Question 0 at the beginning of this lesson. Would you change your answer now that you've completed the lesson? How would you summarize the topic of this lesson now?
4. ☐ What questions do you have about this lesson?

Looking Ahead

1. ☐ Read the opening paragraphs in Lesson 1-4 carefully, then answer Question 0 in preparation for that lesson.

1-3 Applications Name ____________________

1. According to this diagram from Boston Children's Hospital, there are a lot of similarities between the symptoms you experience when you have a cold, and symptoms caused by allergies.

a. List the symptoms that colds and allergies have in common.

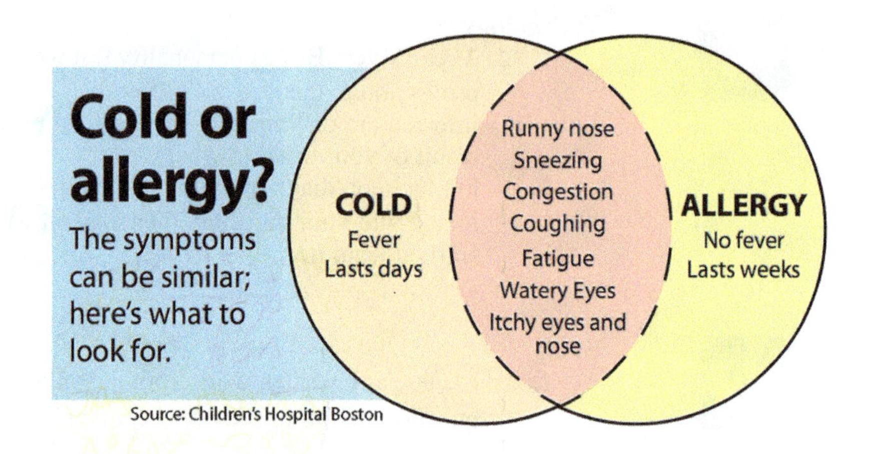

b. List the symptoms that help you tell the difference between a cold and allergies.

2. In a survey of 85 college students, 72 use Facebook, 31 use Google+, and 21 use both. Use this information to fill in the Venn diagram to the right, then use it to answer the questions.

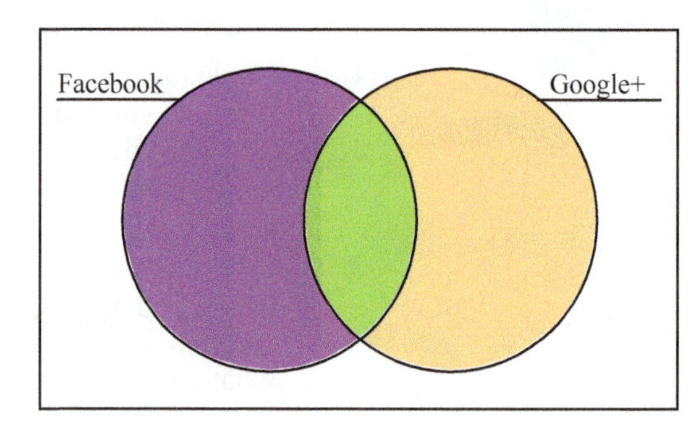

a. How many use Google+ only?

b. How many use Facebook only?

c. How many use neither?

1-3 Applications Name ______________________________

Two hundred patients suffering from depression enrolled in a clinical trial to test the effects of various antidepressants. The Venn diagram shows the number of patients who were given one or more of the three drugs in the trial. Use the diagram to answer Questions 3–8.

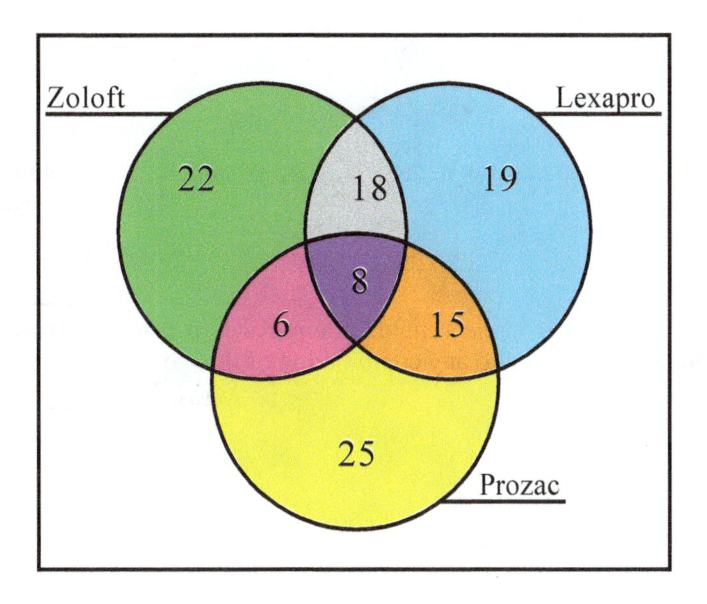

3. How many patients were given all three drugs?

4. How many patients were given just one drug?

5. How many patients were given Lexapro and Prozac but not Zoloft?

6. How many patients were given exactly two of the three drugs?

7. How many patients were given at least Zoloft and Prozac?

8. How many patients were given a placebo containing none of those three drugs?

Optional Critical Thinking Activity

As mentioned in the math note on page 21, it's possible to draw a Venn diagram with four circles.

1. Count the number of different regions in a Venn diagram with two circles, and in one with three circles. Include the portion inside the rectangle but outside of the circles.

2. Based on your answers to Question 1, make a guess as to how many regions you think there will be in a Venn diagram with four circles.

3. In the space below, try to draw a four-circle Venn diagram. Keep in mind that there will need to be a region inside all four circles, a region common to any combination of three of the four, and a region common to any pair.

4. Was your educated guess in Question 2 right?

Lesson 1-4 Take a Guess!

Learning Objectives

- ☐ 1. Make educated guesses.
- ☐ 2. Plot points on a number line.
- ☐ 3. Compare numbers using inequality symbols.
- ☐ 4. Approximate square roots.

Don't be afraid to fail. Be afraid not to try.
– Michael Jordan

Consider this scenario: you're a contestant on a game show, and you're given a trivia question worth $50,000. Pressure! The question is "Which state was first to secede from the union in 1860?" Do you know the answer? If so, terrific – drinks are on you. But if you don't know the answer, you'd have to guess. Would your list of guesses include the Pacific Ocean, Justin Bieber, Twinkies, and a goat? Of course not. Those would be completely silly.

The point is that "guessing" doesn't usually mean just throwing out some random answer and praying for a miracle. Instead, we use information we know to be true and some reasoning ability to make an **educated guess.** In this section, we'll try to gain some confidence with educated guessing, which sometimes goes by a different name in math: **estimation.**

From an instructor's point of view, it seems like in math classes, a lot of students are afraid to make educated guesses. But sometimes that's a completely reasonable step in problem solving. Nobody likes to be wrong, but if you're not willing to at least answer a question, you have no chance at ever being right. Not being afraid of failure is one of the characteristics shared by almost everyone that's very successful in their chosen field. So let's overcome the fear of being wrong, and build the confidence to take a guess.

0. After reading the opening paragraphs, what do you think the main topic of this section will be?

1-4 Class

An **isotherm** is a curve on a map that connects locations where the temperature is the same. By looking at a large map with isotherms, you can get a rough idea of what the temperature is like in many different locations. Using the map below, estimate the temperature in each city.

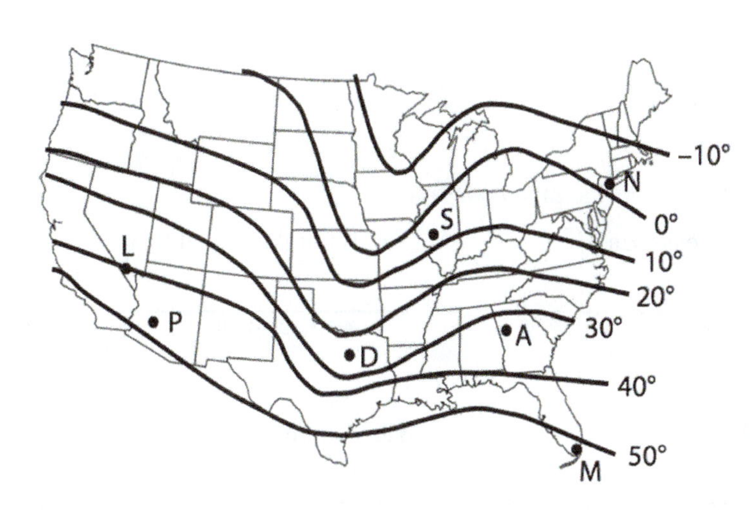

1. Las Vegas, NV (L) 40

2. Phoenix, AZ (P) 48

3. Dallas, TX (D) 25

4. Springfield, IL (S) 8

5. Atlanta, GA (A) 33

6. Miami, FL (M) 52

7. New York, NY (N) -2

Estimation is a hugely valuable skill when it comes to reading information from charts and graphs. In fact, we used that skill (without even realizing it!) in Lesson 1-2 when getting information from bar graphs. When the length of a bar falls between two lines, like in the graph on Page 11, we have to estimate its height. We can practice this useful skill by looking at number lines.

8. Use the number line to estimate the value represented by each point. Write a short explanation of how you decided on your answer.

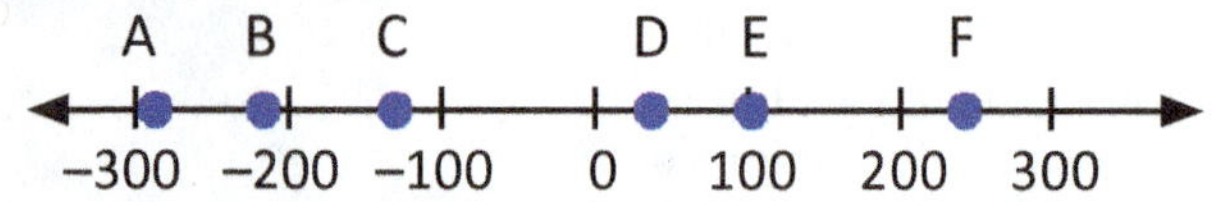

Point	Value	Explanation
A	-285	THE VALUE IS APPROX 85% BELOW -300 MARK
B	-215	THE VALUE IS APROX 15% ABOVE -200 "
C	-125	" " " " 25% " -100 "
D	30	" " " " 30% " 0 "
E	102	" " " JUST ABOUT THE 100 MARK
F	233	" " " APROX 33% ABOVE THE 200 MARK

You might have started Question 8 by thinking "Point B is a little bit left of –200." We can express the fact that point B is to the left of –200 in symbols by writing B < –200. This is read as "B is less than negative 200."

Use the next number line to answer Questions 9 and 10.

-5 -4 -3 -2 -1 0 1 2 3 4 5

9. We write –2 < 3 because –2 is to the LEFT of 3.

10. We write 3 > –2 because 3 is to the RIGHT of –2.

Math Note

The expression A > –300, read as "A is greater than negative 300," expresses the fact that A represents a number to the right of –300 on a number line.

Insert <, =, or > in the blank to make each statement true.

11. –10 > –15

12. –100 < 50

13. $\sqrt{25}$ = 5

For Questions 14–19, plot a point on the number line that meets the given condition.

14. A number greater than 3

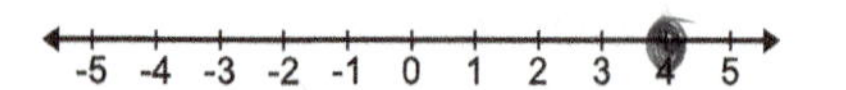

15. A number less than 3

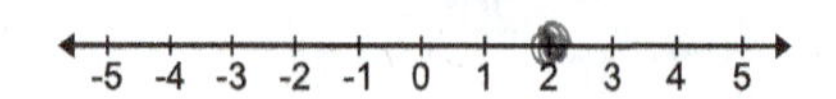

16. A number equal to 3

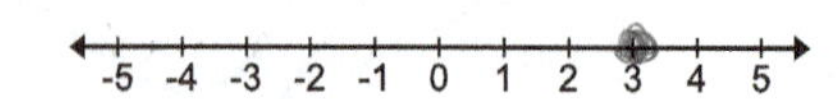

17. A number greater than –2

18. A number less than –2

-5 -4 -3 -2 -1 0 1 2 3 4 5

19. A number equal to –2

-5 -4 -3 -2 -1 0 1 2 3 4 5

Estimation is commonly used when dealing with square roots. We know, for example, that $\sqrt{49} = 7$. But what is $\sqrt{47}$? Since 47 is a little smaller than 49, it seems reasonable to estimate that $\sqrt{47}$ is a little less than 7. To get more accurate (but still approximate) values for square roots, we can use a calculator or a spreadsheet.

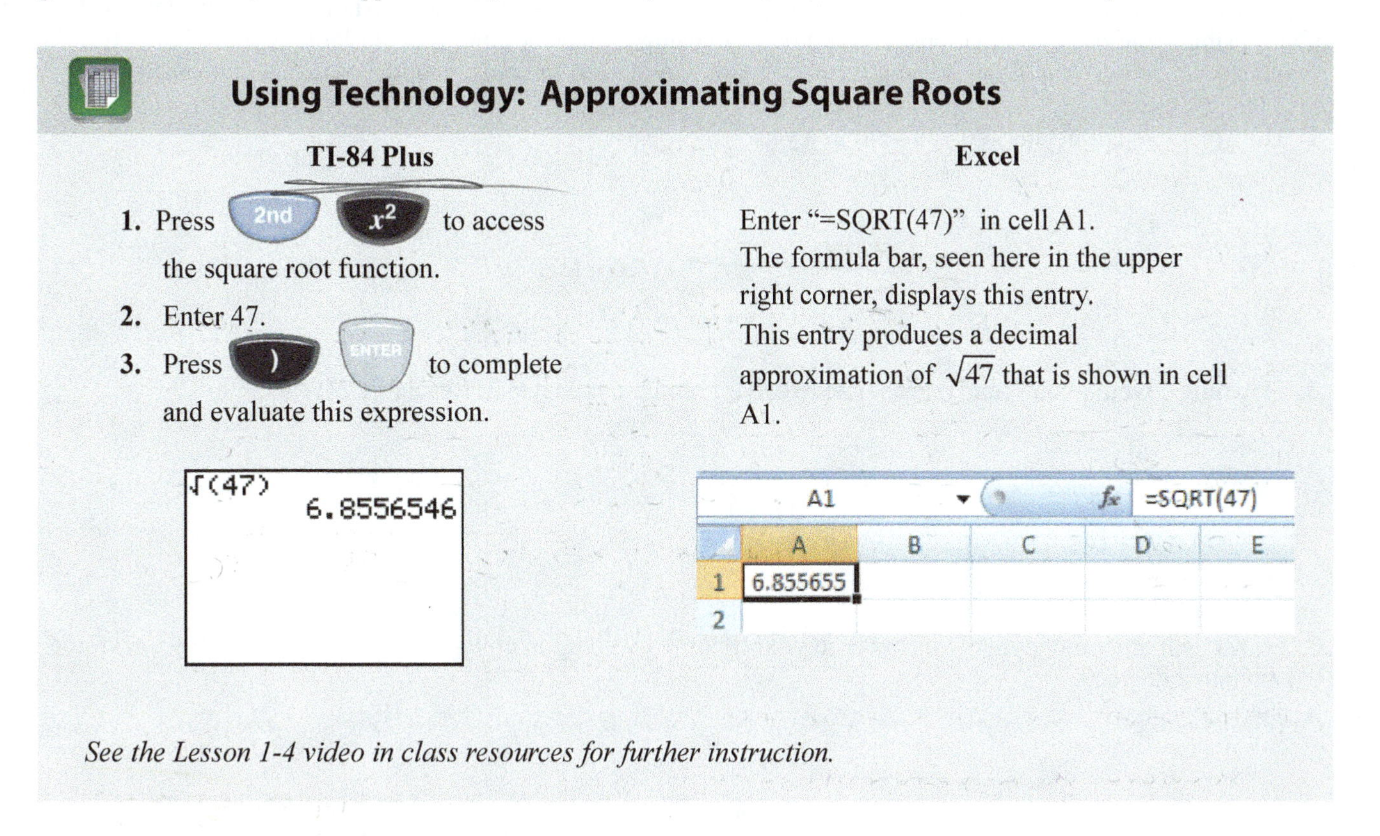

20. Complete the following table of common square roots.

Fill in the correct value to make each equation true. No calculators!		Estimate each square root to the nearest whole number and fill in the relationship between the square root and your estimate with either < or >.	Use a calculator or a spreadsheet to approximate the following square roots to the nearest hundredth.
$\sqrt{1} = 1$	$\sqrt{64} = 8$	$\sqrt{11} > 3$	$\sqrt{11} \approx$ = 3,3166 = 3.32
$\sqrt{4} = 2$	$\sqrt{81} = 9$		
$\sqrt{9} = 3$	$\sqrt{100} = 10$	$\sqrt{78} < 9$	$\sqrt{78} \approx$ = 8,831 = 8.83
$\sqrt{16} = 4$	$\sqrt{121} = 11$		
$\sqrt{25} = 5$	$\sqrt{144} = 12$	$\sqrt{99} < 10$	$\sqrt{99} \approx$ = 9.949 ≈ 9.94
$\sqrt{36} = 6$	$\sqrt{169} = 13$		
$\sqrt{49} = 7$	$\sqrt{196} = 14$	$\sqrt{124} < 11$	$\sqrt{124} \approx$ = 11,1355 ≈ 11.14
	$\sqrt{225} = 15$		

1-4 Group

Spend some time thinking about each question below. After talking with your group, make your best guess and explain your reasoning. Remember, these are supposed to be EDUCATED guesses, not shots in the dark, so think about using some basic calculations to help you. No using phones or computers to look anything up yet! After making your guess, write down some information you would need to know to find a more accurate answer.

1. How many text messages do you send in a year?

TXT PER DAY 5
DAYS IN A YEAR X 365

1825 I SEND APROX 1825 PER YEAR

2. How much would you spend on gasoline to drive from here to Daytona Beach, FL?

PORTLAND OR. → DATONA BEACH FL. 3000
VEHICLE GETS 25 MILES PER GALLON 25 = 3000 / 25 = 120
120 GALLONS X $3.00 PER GALLON OF GAS = 120 X $3 = $360

3. How much more money will you make in your lifetime by going to college instead of entering the workforce without a degree?

CURRENT EARNING PER YEAR $60,000
EARNING PER YEAR AFTER DEGREE. $30,00
I HAVE APROX 11 YEARS TILL RETIREMENT
I WOULD LOSE APROX $330,000 IN MY LIFE TIME

$60,000
$30,00
$30000
X 11 = 330,000

4. What would your car payment be on a $10,000 car if you were going to pay it off in 3 years?

12 MONTH IN A YEAR X 3 YEARS = 36 MONTH
CAR LOAN IS $10,000 / 36 = 277.77 = 278 PER MONTH

5. How much time do you spend on Facebook and other social media sites each week?

I AVERAGED 3 MIN'S PER DAY ON FACEBOOK X 7 DAYS PER WEEK

3
X7
21

I SPEND APROX 21 MINS PER DAY ON FACEBOOK

6. How much time do you spend watching TV each week?

HOURS PER DAY WATCHING T.V. = 3
DAYS IN A WEEK = X 7
21

I SPEND APROX 21 HOURS PER WEEK WATCHING T.V.

7. How many people in the United States do not smoke cigarettes?

THERE IS APROX 318 MILLION PEOPLE IN THE US
APROX .65% DO NOT SMOKE

318
X.65
206

APROX 206 MILLION PEOPLE IN THE U.S. DO NOT SMOKE

8. How much money do you spend on soda or coffee each week?

I SPEND APROX $2.00 PER DAY ON SODA
THERE IS 7 DAY'S PER WEEK

7
X2
= 14

I SPEND APROX $14 PER WEEK ON COFFEE AND SODA

9. If you put $100 into an Individual Retirement Account (IRA) today, how much would it be worth when you retire?

6% INTREST 100 X .06 = 6 X 11 YEARS = $66 + 100 = $166
I FOUND AN IRA AT 6% THEN MULTIPLIED IT X $100 TO FIND MY INTEREST FOR EACH YEAR THEN MULTIPLIED IT TIMES 11 YEARS AND ADDED IT TO MY $100
MY IRA WILL BE WORTH $166

10. How much money do you waste when you skip one class?

I HAVE 5 CLASSES THE COMBINED AMOUNT OF THOSE CLASSES IS APROX $1,200, THERE ARE 20 CLASSES EACH TERM

$1,200
20
= $60

I WASTE APROX $60 EACH TIME I SKIP CLASS

11. How many people in your state are on their cell phone right now?

THE POPULATION OF OREGON IS APROX 4 MILLION
APROX .5% IS ON THE PHONE

4,000,000
X .5% =

4 M
X.005 = 20,000

12. How tall is a stack of 1,000,000 dollar bills? *(See a cool illustration in Lesson 1-4 resources online.)*

A STACK OF BILLS IS APROX .4 INCHES
I MULTIPLIED 1,000,000 X .4

100 ⟌ 1,000,000 = 10,000 X .4 = 4,000 12 ⟌ 4000 = 333.330

400,000 INCHES
12 INCHES PER FOOT 12 ⟌ 400000 = 33,333 FEET TALL

A STACK OF 1,000,000 DOLLAR BILLS IS APROXIMETLY 33,330 FEET HIGH

1-4 Portfolio

Name ______________________________

Check each box when you've completed the task. Remember that your instructor will want you to turn in the portfolio pages you create.

Technology

1. ☐ Use Excel to make a table of square roots for all of the whole numbers from 1 to 25. Format so the results are rounded to two decimal places. A template to help you get started can be found in the online resources for this lesson; also see the technology video under the resources for Lesson 1-4 for help.

Skills

1. ☐ Include any written work from the online skills assignment along with any notes or questions about this lesson's content.

Applications

1. ☐ Complete the applications problems.

Reflections

Type a short answer to each question.

1. ☐ What kinds of things would you think about when making an educated guess in a math problem?
2. ☐ Describe at least one situation where estimating is good enough, and one where it isn't.
3. ☐ When using a calculator or spreadsheet to find square roots, why did we say "approximate" rather than compute? An Internet search will help.
4. ☐ Take another look at your answer to Question 0 at the beginning of this lesson. Would you change your answer now that you've completed the lesson? How would you summarize the topic of this lesson now?
5. ☐ What questions do you have about this lesson?

Looking Ahead

1. ☐ Read the opening paragraph in Lesson 1-5 carefully, then answer Question 0 in preparation for that lesson.

1-4 Applications

Name WILLIAM BEELER

1. Pie charts can be used to describe many things – including pies! A small bakery shop in Hamilton, OH sells an average of 120 pies per month, with the types broken down in the given pie chart. Use the chart to make an educated guess as to the number of each type the shop sold. Explain your reasoning for each estimate.

Type	Number
Apple	58
Other	7
Pumpkin	22
Strawberry	7
Cherry	26

I VISUALLY ESTIMATED THAT APPLE IS NEARLY HALF OF PIE GRAPH SO I USED 48% IN MY EQUATION AND SO ON

Other
Pumpkin
Strawberry
Apple
Cherry

Reasoning: TOTAL # OF PIES SOLD OF 120

APPLE IS APROX 48% = .48 X 120 = 58
OTHER " " 6% = .06 X 120 = 7
PUMPKIN " " 18% = .18 X 120 = 21
STRAWBERRY " 6% = .06 X 120 = 7
CHERRY " " 22% = .22 X 120 = 26
100

2. Use the map to estimate the distance from St. Louis to New York, and explain your reasoning.

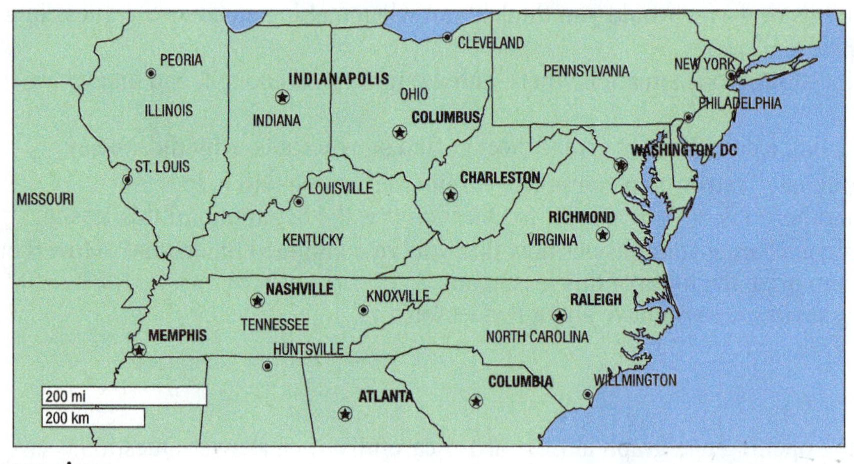

Distance estimate:

1 How?

Reasoning:

I ESTIMATED IT TO BE APROX 4.5 X THE 200 MILE BAR GRAPH

200
x 4.5
= 900 = THE DISTANCE BETWEEN ST. LOUIS TO NEW YORK IS APROX 900 MILES

1-4 Applications

Name ____________________

Shopping provides a really good example of the value of estimation. Most people have a rough idea of how much they have to spend in a given shopping trip. While you could pull out the old cell phone calculator and add up all of your purchases to the cent, who has time for that? Instead, it's helpful to round prices to easy-to-use numbers and keep rough track of the total. Use this idea to estimate the total cost of each group of items in Questions 3–5.

3. Four boxes of mac and cheese (89 cents per box)
A gallon of orange juice ($3.79)
A loaf of bread ($1.79)
Two pounds of ground turkey ($2.97 per pound)

4. Two pairs of jeans ($24.95 and $32.95)
A three-pack of socks ($7.99)
Three workout shirts ($11.99 each)

5. Two Xbox games (regular price $29.50 each, on sale for 50% off)
iPhone case ($18.95)
Headphones ($22.49)

6. Estimation is useful for tipping when you're dining out, too. If the service is good, 20% of the overall bill is a nice tip to leave. Here's a clever way to find that: move the decimal of the total amount one place to the left: that gives you 10%. Round this value to the nearest dollar and double the amount, and voila! Twenty percent. Use this to find a reasonable tip for good service on each bill.

a. $31.45

b. $129.80

Lesson 1-5 Do You Have Anything To Add?

Learning Objectives

- ☐ 1. Identify when it's appropriate to add quantities.
- ☐ 2. Practice addition and subtraction.
- ☐ 3. Interpret bank statements.
- ☐ 4. Use Excel to compute sums.
- ☐ 5. Calculate the perimeter of rectangles and triangles.

We are what we repeatedly do. Excellence, therefore, is not an act, but a habit.
– Aristotle

We've all heard statistics describing the collective weight problem that we as Americans have. So it's not at all surprising that counting calories has become something of a national pastime. Nutritional information labels are now mandated by law on food packaging, and more and more restaurants are including calorie and fat information on their menus. In order to keep track of what you're putting in your body, you need to be able to add up amounts of calories, fat, carbohydrates, protein, and others. This requires identifying and adding the amounts that correspond to like ingredients. This is a skill that will come in handy throughout this course (and beyond).

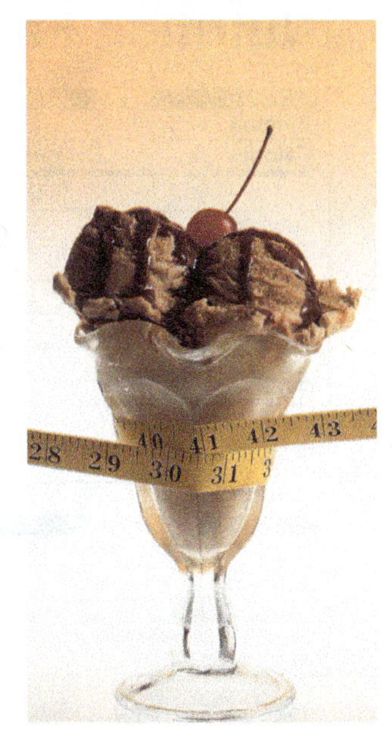

0. After reading the opening paragraph, what do you think the main topic of this section will be?

1-5 Group

My coauthor's daughter Charlotte is in kindergarten, and her favorite lunch consists of a peanut butter and jelly sandwich, carrot slices, yogurt, and juice. The nutrition facts for the food in her lunch are given here. Use these facts to answer questions that follow. A sandwich is made from two servings of bread, ½ serving of peanut butter, and ½ serving of jelly. Charlotte also gets one serving each of carrots and yogurt along with one juice box.

Wheat Bread

Nutrition Facts Serving Size 25g	
Amount Per Serving	
Calories 66	Calories from Fat 8
	% Daily Value*
Total Fat 1g	1%
Saturated Fat 0g	1%
Trans Fat 0g	
Cholesterol 0mg	0%
Sodium 130mg	5%
Total Carbohydrate 12g	4%
Dietary Fiber 1g	4%
Sugars 1g	
Protein 3g	
Vitamin A 0% • Vitamin C	0%
Calcium 4% • Iron	5%

*Percent Daily values are based on a 2,000 calorie diet. Your daily values may be higher or lower depending on your calorie needs.

Peanut Butter

Nutrition Facts Serving Size 32g	
Amount Per Serving	
Calories 188	Calories from Fat 135
	% Daily Value*
Total Fat 16g	25%
Saturated Fat 3g	17%
Trans Fat 0g	
Cholesterol 0mg	0%
Sodium 147mg	6%
Total Carbohydrate 6g	2%
Dietary Fiber 2g	8%
Sugars 3g	
Protein 8g	
Vitamin A 0% • Vitamin C	0%
Calcium 1% • Iron	3%

*Percent Daily values are based on a 2,000 calorie diet. Your daily values may be higher or lower depending on your calorie needs.

Strawberry Jelly

Nutrition Facts Serving Size 21g	
Amount Per Serving	
Calories 56	Calories from Fat 0
	% Daily Value*
Total Fat 0 g	0%
Saturated Fat 0g	0%
Trans Fat 0g	
Cholesterol 0mg	0%
Sodium 6mg	0%
Total Carbohydrate 15g	5%
Dietary Fiber 0g	1%
Sugars 11g	
Protein 0g	
Vitamin A 0% • Vitamin C	0%
Calcium 0% • Iron	0%

*Percent Daily values are based on a 2,000 calorie diet. Your daily values may be higher or lower depending on your calorie needs.

Carrots

Nutrition Facts	
Serving Size 61g	
Amount Per Serving	
Calories 25	Calories from Fat 1
	% Daily Value*
Total Fat 0 g	0%
Saturated Fat 0g	0%
Trans Fat 0g	
Cholesterol 0mg	0%
Sodium 42mg	2%
Total Carbohydrate 6g	2%
Dietary Fiber 2g	7%
Sugars 3g	
Protein 1g	
Vitamin A 204% • Vitamin C	6%
Calcium 2% • Iron	1%

*Percent Daily values are based on a 2,000 calorie diet. Your daily values may be higher or lower depending on your calorie needs.

Yogurt

Nutrition Facts	
Serving Size 125g	
Amount Per Serving	
Calories 119	Calories from Fat 2
	% Daily Value*
Total Fat 0 g	0%
Saturated Fat 0g	1%
Trans Fat 0g	
Cholesterol 2mg	1%
Sodium 72mg	3%
Total Carbohydrate 24g	8%
Dietary Fiber 0g	0%
Sugars 24g	
Protein 6g	
Vitamin A 0% • Vitamin C	1%
Calcium 19% • Iron	0%

*Percent Daily values are based on a 2,000 calorie diet. Your daily values may be higher or lower depending on your calorie needs.

Juice Box

Nutrition Facts	
Serving Size 1 Juice Box	
Amount Per Serving	
Calories 60	Calories from Fat 0
	% Daily Value*
Total Fat 0 g	0%
Saturated Fat 0g	0%
Trans Fat 0g	
Cholesterol 0mg	0%
Sodium 10mg	1%
Total Carbohydrate 15g	5%
Dietary Fiber 0g	0%
Sugars 14g	
Protein 0g	
Vitamin A 10% • Vitamin C	100%
Calcium 10% • Iron	0%

*Percent Daily values are based on a 2,000 calorie diet. Your daily values may be higher or lower depending on your calorie needs.

Find the total amount of each in Charlotte's lunch.

1. Total calories

BREAD = 66 X 2 = 132
PB = 188 X .5 = 94
JELLY = 56 X .5 = 28
CARROT = 25 X 1 = 25
YOGURT = 119 X 1 = 119
JUICE = 60 X 1 = 60
= 458 MG TOTAL CALORIES

2. Total calories from fat

BREAD = 8 X 2 = 16
PB = 135 X .5 = 67.5
JELLY = 0 X 1 = 0
CARROTS = 1 X 1 = 1
YOGURT = 2 X 1 = 2
JUICE = 0 X 1 = 0
TOTAL CALORIES FROM FAT = 86.5 MG

3. Total milligrams (mg) of sodium

BREAD = 130 X 2 = 260
PB = 147 X .5 = 73.5
JELLY = 6 X .5 = 3
CARROT = 42 X 1 = 42
YOGURT = 72 X 1 = 72
JUICE = 10 X 1 = 10
SODIUM = 460.5 MG

4. Total grams (g) of carbohydrates

BREAD = 12 X 2 = 24
P.B = 6 X .5 = 3
JELLY = 15 X .5 = 7.5
CARROT = 6 X 1 = 6
YOGURT = 24 X 1 = 24
JUICE = 15 X 1 = 15
TOTAL CARBS = 79.5 G

5. Total grams (g) of sugars

BREAD = 1 X 2 = 2
PB = 3 X .5 = 1.5
JELLY = 11 X .5 = 5.5
CARROT = 3 X 1 = 3
YOGURT = 24 X 1 = 24
JUICE = 14 X 1 = 14
SUGARS = 50 G

6. Total milligrams (mg) of cholesterol

BREAD = 0 X 2 = 0
PB = 0 X .5 = 0
JELLY = 0 X .5 = 0
CARROT = 0 X 1 = 0
YOGURT = 2 X 1 = 2
JUICE = 0 X 1 = 0
TOTAL CHOLEST = 2 MG

7. If we use the letter P to stand for the phrase "grams of protein", what is the result of the calculation below, and what is the significance of the answer?

$6P + 4P + 0P + 1P + 6P + 0P$ = 17 GRAMS OF PROTEIN

1-5 Class

Here are some addition problems for you to work out. Some can be done very quickly; some need a little more thought; others can't be done at all. Perform each addition that you can. For those that can't be done, explain why.

1. $\frac{2}{7}+\frac{3}{7}$

2. $0.5 + 0.2$

3. 6 mi + 8 mi

4. 7 g protein + 15 g protein

5. $6\sqrt{2}+5\sqrt{2}$

6. $4x + 5x$

7. \$2 + 23¢

8. 8 min + 30 sec

9. 3 days + 5 hours

10. $\frac{2}{5}+\frac{3}{7}$

11. $0.2 + 0.03$

12. 10g sodium + 17 g carbs

13. 5 hr + 3 mi

14. 6 min + \$5

15. $3y+7y^2$

16. Before adding quantities, make sure that they are

____________________ ____________________.

Math Note

You probably recognize addition problems like numbers 6 and 15 from the study of **algebra**, where we use symbols (often letters) called **variables** to represent quantities that can VARY.

1-5 Group (Again)

A bank statement gives you a detailed look at the activity in your account over a period of time. All you need to know to understand a bank statement is a few words of vocabulary, and how to add and subtract positive and negative numbers. Use the sample statement below to answer each question. (Amounts in parentheses are negative.)

Bank Statement

Statement date			Previous balance	$2,358.25
5/17/12			4 credits	$1,007.13
			11 debits	($1,000.66)
			Ending balance	$2,364.72

Date	Description	Credits	Debits	Balance
4/17/12	Rent		($400.00)	$1,958.25
4/19/12	Gas		($44.33)	$1,913.92
4/20/12	Paycheck	$479.68		$2,393.60
4/23/12	Groceries		($79.24)	$2,314.36
4/24/12	New clothes		($57.18)	$2,257.18
4/26/12	Return shirt	$22.01		$2,279.19
5/2/12	Cell phone bill		($68.44)	$2,210.75
5/4/12	Pizza		($11.02)	$2,199.73
5/4/12	Paycheck	$504.88		$2,704.61
5/8/12	Withdraw cash		($200.00)	$2,504.61
5/12/12	Check #278		($40.00)	$2,464.61
5/13/12	Groceries		($48.23)	$2,416.38
5/15/12	Gas		($38.68)	$2,377.70
5/16/12	Pizza		($13.54)	$2,364.16
5/17/12	Interest	$0.56		$2,364.72

1. What does a credit refer to?

AN ADDITION TO YOUR ACCOUND

2. What does a debit refer to?

A SUBSTRACTION FROM YOUR ACCOUNT

3. What is meant by a balance?

THE CURRENT TOTAL AMOUNT IN THE ACCOUNT

4. What would the ending balance be with an additional credit of $100?

$2,364.72 + $100.00 = $2,464.72
THE ENDING BALANCE WOULD BE $2,464.72

5. What would the ending balance be with an additional debit of $50?

END BAL $2,464.72
A DEDUCTION OF $50.00
= $2,414.72

6. What would the ending balance have been if the account holder didn't return the shirt on 4/26?

= $2,414.72
− 22.01
$2,392.71

7. What would the ending balance be if we remove the $200 cash withdrawal from 5/8?

$2,392.71
+ 200.–
2,592.71

At some point in elementary school, you were taught to remember rules for adding and subtracting positive and negative numbers. But did you ever try to understand why those rules make sense? In previous classes, you might have been so focused on *remembering* rules that you never really *thought* about them. Our goal is to *understand* why rules make sense. Let's see if our study of bank statements can shed some light on this important subject.

8. Credits add to the balance in an account, while debits subtract from it. What is the result of adding two credits to the overall balance? What can you conclude about adding two positive numbers?

THE ACTUAL BAL WILL INCREASE

ADDING 2 POSITIVE NUMBERS WILL INCREASE YOUR BALANCE

9. What's the result of adding two debits to the overall balance? What can you conclude about adding two negative numbers?

THE BALANCE WOULD DECRESE

ADDING 2 NEGATIVE NUMBERS WILL DECREASE YOUR BALANCE

10. What happens to the balance if you add a credit, then a debit? What can you conclude about adding a positive and a negative number?

THE BALANCE WOULD INCREASE THEN DECREASE

YOUR BAL WILL CHANGE WITH EACH DEBIT AND CREDIT

11. Removing a debit is in effect subtracting a negative amount from the balance. How does that affect the account balance? What does this tell you about subtracting a negative number?

REMOVING A DEBIT WILL INCREASE YOUR BALANCE,

SUBTRACTING A NEGATIVE NUMBER HAS THE SAME RESULT AS ADDING A CREDIT

12. If your balance is \$1,150 and you deposit \$200, is the result the same as if your balance is \$200 and you deposit \$1,150? What does that tell you about order and addition?

YES! THE ORDER OF ADDITIONS HAS NO EFFECT ON THE END BALANCE

13. If your balance is \$700 and you have a debit of \$250, is the result the same as if your balance is \$250 and you have a debit of \$700? What does that tell you about order and subtraction?

YES! THE ORDER OF SUBTRACTIONS HAS NO EFFECT ON THE END BALANCE

14. Suppose you have three checks to deposit. The amounts are \$22.37, \$75.00, and \$125.00. When filling out the deposit slip, one way to add the amounts is to first add \$22.37 and \$75.00 to get \$97.37. Then add \$97.37 and \$125.00 to obtain the total of \$222.37. Find a similar, but easier approach that gets the same result. What does this tell you about addition?

I FIND IT EASIER TO ADD ALL THREE DEPOSITS AT ONCE AS I FIND IT TO BE ONE LESS STEP.

\$22.37
75.00
+125.00
= 222.37

IT DOES NOT MATTER HOW YOU ADD ALL DEPOSITS

15. List some words or phrases that are used to indicate that addition should be done.

I WILL MAKE DEPOSIT'S TO THIS ACCOUNT

I WILL RETURN AN ITEM THAT IS A DEBIT

I DEPOSITED MY PAYCHECK

16. List some words or phrases that are used to indicate that subtraction should be done.

I WITHDREW MONEY FROM MY ACCOUNT

I PAID THE RENT

I ORDERED A PIZZA TO BE DELIVERED

Spreadsheets are really good at doing large calculations that are inconvenient to do by hand. If you have to add three credits to a balance, a hand calculation or using a calculator are both perfectly fine. But if you have to add 40 debits over the course of a month, a spreadsheet will be a huge help.

You can use the "+" symbol for addition, but the SUM command is more useful. For example, if you enter

=A1+A3+A4 or =SUM(A1,A3,A4)

each of these will add the values in cells A1, A3, and A4. But to add an entire range of cells, you can enter

=SUM(A1:A10)

which will add all of the values in the cells from A1, A2, A3, and so on, down to A10. = SUM(A1:A10

Using Technology: The SUM Command in Excel

To add values in neighboring rows or columns in Excel:

1. Enter the values either into a row or a column. In this case, values have been entered into cells A1, A2, A3, A4, and A5.
2. To compute the sum of these 5 values and display that value in cell A6, enter "=SUM(A1:A5)" in cell A6.
3. This is equivalent to entering "=A1 + A2 + A3 + A4 + A5".

Note that the cells have to be consecutive to use this command.

See the Lesson 1-5 video in class resources for further information.

A6 ▾ fx =SUM(A1:A5)

	A	B	C
1	5		
2	7		
3	8		
4	12		
5	3		
6	35		
7			

1-5 Portfolio

Name WILLIAM BERGER

Check each box when you've completed the task. Remember that your instructor will want you to turn in the portfolio pages you create.

Technology

1. ☐ Design a spreadsheet to find the ending balance for the bank statement on page 40. If you enter the beginning balance and credits as positive numbers and debits as negative numbers, you can use the SUM command in one column. If you want to try something fancier, you can put the credits in one column, the debits in another, then sum each column and decide how to use each to calculate the ending balance. A template to help you get started can be found in the online resources for this lesson.

Skills

1. ☐ Include any written work from the online skills assignment along with any notes or questions about this lesson's content.

Applications

1. ☐ Complete the applications problems.

Reflections

Type a short answer to each question.

1. ☐ In your own words, describe the importance of like terms in addition.
2. ☐ Analyze a recent bank statement for your own account, or one for a family member. Describe the significance of the credits and debits.
3. ☐ Take another look at your answer to Question 0 at the beginning of this lesson. Would you change your answer now that you've completed the lesson? How would you summarize the topic of this lesson now?
4. ☐ What questions do you have about this lesson?

Looking Ahead

1. ☐ Read the opening paragraph in Lesson 1-6 carefully, then answer Question 0 in preparation for that lesson.

1-5 Applications

Name WILLIAM BEEZER

10/10

1. Bob and Tom left a town at the same time. One drove 200 miles east. The other drove 180 miles west. What's the distance between the two cars? I WILL ADD 200+180=380
THE DISTANCE BETWEEN THE TWO CARS IS 380 miles
200 + 180 = 380 MILES

2. Chick and Kristi left a town at the same time. One drove 300 miles east. The other drove 180 miles east. What's the distance between the two cars? I WILL SUBTRACT 180 FROM 300
300 − 180 = 120 THIS DISTANCE BETWEEN THE TWO CARS IS 120 miles

The **perimeter** of a figure is found by taking the sum of the lengths of each side of that figure. Find the perimeter of each figure.

3.

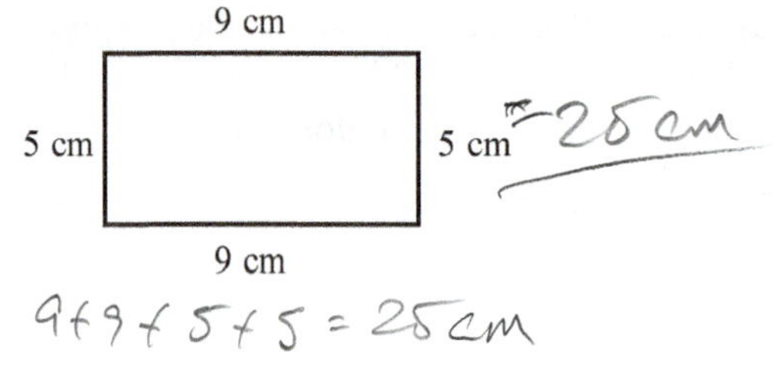

= 25 cm

9+9+5+5 = 25 cm

4.

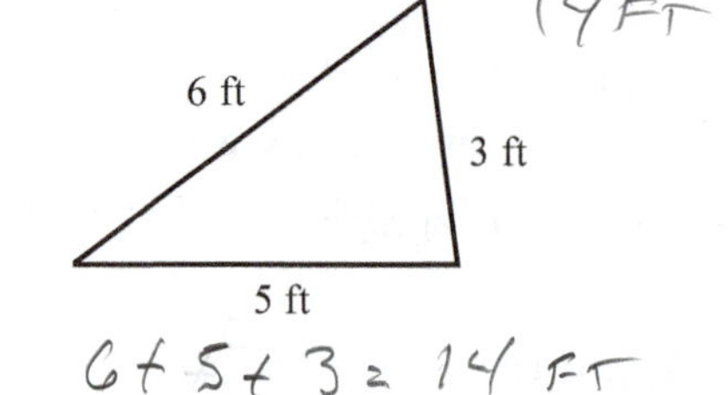

14 FT

6+5+3 = 14 FT

5. Find the length of the unlabeled side if the perimeter is 24 in. 4+11=15 − 4 24 − 15 = 29 IN

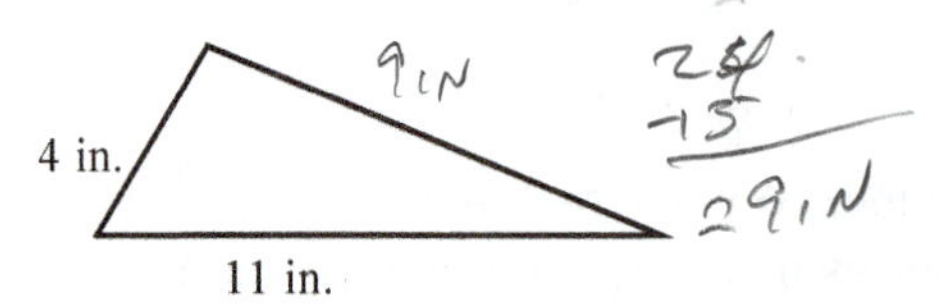

6. Find the length of the unlabeled sides if the perimeter is 40 m. 8+8=16 16 − 40 = 24 24 x .5 = 12

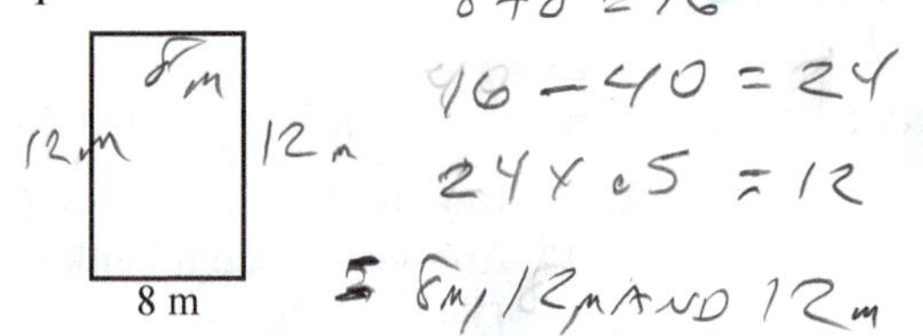

8m, 12m AND 12m

Answer the following questions about the spreadsheet, which shows the points earned by a student on a 10-question quiz worth a total of 50 points.

C12 =SUM(C2:C11)

	A	B	C
1	Problem #	Possible	Points
2	1	6	5
3	2	7	7
4	3	4	3
5	4	5	5
6	5	9	8
7	6	3	3
8	7	3	3
9	8	3	0
10	9	2	2
11	10	8	6
12		50	42
13			

7. What percent of possible points did this student earn?
42/50 = 50 X .42 = .84 = 84%
THE STUDENT EARNED 84% OF THE POSSIBLE POINTS

8. Write down all of the numbers that were added to calculate the value in cell C12.
5, 7, 3, 5, 8, 3, 3, 2, 6

9. A formula was entered in cell B12, not the number 50. Write the formula that was entered.
=SUM(B2:B11)

Lesson 1-6 It's About Accumulation

Learning Objectives

- ☐ 1. Consider areas of your life where many small events add up to one large event.
- ☐ 2. Interpret multiplication as repeated addition.
- ☐ 3. Refresh multiplication and division skills.

A journey of a thousand miles begins with a single step.
– Lao-Tzu

The best time to plant a tree is twenty years ago. The second best time is now.
– Chinese proverb

In my experience as both a student and a professor, maybe the single biggest obstacle to succeeding in college is procrastination. As the work you have to do mounts, it's easy to feel overwhelmed at times. We're all familiar with being faced with so much to do that we just don't know where to start. Unfortunately, too many people take the default path, which is to do nothing constructive and put the important stuff off. Sometimes, though, you just have to force yourself to take that first step on the thousand mile journey. The cumulative effect of doing small bits of work can accomplish a lot!

0. After reading the opening paragraph, what do you think the main topic of this section will be?

1-6 Class

1. Write about a time you put off something that you had to do, and then regretted it later.

I PUT OFF DOING THE DISHES, THEN A BUDDY OF MINE SHOWED UP UNEXPECTEDLY WITH HIS WIFE AND HER GIRLFRIEND

2. Write down an activity that you're good at, like a sport, playing an instrument, a video game, or something else. How good were you when you started, compared to now? How much time went into building your skill at that activity? I AM GOOD AT WATCHING T.V. WHEN I STARTED I HAD TO GET UP, OUT OF THE CHAIR AND WALK TO THE T.V. TO CHANGE THE CHANNEL OR VOLUME. NOW I AM THE MASTER OF MY CABLE BOX REMOTE

3. Write about a task you've performed that required many different small steps to accomplish something bigger.

4. What do you do when you're overwhelmed with work and need to develop a plan?

I WRITE A LIST IN A COLLOM ON THE LEFT SIDE OF THE PAGE, THEN I

1-6 Group

Some of the problems in this activity require calculations. Feel free to use a calculator or computer, but make sure you write down enough work to explain your results.

1. My friend Charles noticed that one of his students came to class every day with two cups of coffee from a well-known coffee chain that isn't exactly famous for their low prices. He asked her if she'd ever thought about how much she spent on that coffee over the course of a year; not surprisingly, she had not. Let's help her out. If the student buys two cups of coffee every day of the year at $3.50 each, how much will she spend in one year? (Remember, two cups per day.)

2 CUPS PER DAY X 365 DAYS PER YEAR = 730 CUPS PER YEAR

365 × 2 = 730

730 × $3.50 FOR EACH CUP OF COFFEE = $2,555

SHE SPENT $2,555 PER YEAR

2. There are two ways to calculate the amount spent on coffee in Question 1. First, you can add $7.00 + $7.00 + $7.00 + $7.00 + ... until you've written $7.00 ___365___ times. Second, you can multiply ___$3.50 PER CUP___ by ___730 CUPS___.

365 DAYS × 2 CUPS PER DAY

730 CUPS × $3.50 = $2,555

3. Use Question 2 to describe the relationship between multiplication and addition.

THE RELATIONSHIP BETWEEN THEM IS YOU CAN END UP WITH THE SAME ANSWER. MULTIPLICATIONS SEEMS TO BE AN EASIER METHOD

4. Going back to our coffee-buying friend in Question 1, complete the table below, based on buying two $3.50 cups of coffee every day. Don't just write an answer: write the entire calculation.

How much will she spend...

... after 1 day?	1 DAY × 2 CUPS = 2 CUPS × $3.50 = $7.00
... after 3 days?	3 DAYS × 2 CUPS = 6 CUPS × $3.50 = $21.00
... after 10 days?	10 DAYS × 2 CUPS = 20 CUPS × $3.50 = $70.00
... after 180 days?	180 DAYS × 2 CUPS = 360 CUPS × $3.50 = $1,260

5. How much money would this student save over the course of the year if she buys coffee at a different location where it costs $2 per cup?

365 DAY × 2 CUPS PER DAY = 730 CUPS PER YEAR × $3.50 PER CUP = $2,555

365 DAYS × 2 CUPS PER DAY = 730 CUPS PER YEAR × $2.00 PER CUP = $1,460

$2,555 AT $3.50 − $1,460 AT $2.00 = $1,095 DIFFERANCE

THE STUDENT WOULD SAVE $1,095 PER YEAR

6. Tommy (who used to work on the docks) and Gina (who works the diner all day) wisely plan to start saving for retirement. They deposit $1,000 into an account to start out. The plan is to deposit $50 into the account each month for the next 20 years. How much money will they have deposited into the account at the end of 20 years?

$50 EACH WEEK
X 12 MONTHS
= $600 PER YEAR
X 20 YEARS
= $12,000

$12,000 DEPOSIT OVER 20 YEARS
+ $1,000 INITIAL DEPOSIT
= 13,000

TOMMY AND GINA WOULD HAVE $13,000 AT THE END OF 20 YEARS

7. When you're trying to save money, it can help motivate you to calculate how much money accumulates little by little. Help Tommy and Gina with some of these calculations by filling in the table below, assuming that they stick with their plan of depositing $50 each month after starting with the initial deposit of $1,000. Again, include all calculations, not just the answer.

How much will they have saved...

... after 1 month?	$50 PER MONTH X 1 MONTH = $50 + $1,000 = $1,050
... after 1 year?	$50 PER MONTH X 12 MONTHS = $600 + $1,000 = $1,600
... after 5 years?	$50 PER MONTH X 5 YEARS = 5X12 = 60 MONTHS X $50 = $3,000 + $1,000 = $4,000
... after 10 years? (They're halfway there.)	10 YEARS AT 12 MONTHS PER YEAR = 120 MONTHS X $50.00 PER MONTH = $6,000 + $1,000 = $7,000

8. How much more can Tommy and Gina save in 20 years if they deposit $75 each month rather than $50?

12 MONTHS
X $50 PER MONTH
$600 PER YEAR
X 20 YEARS
$12,000

12 MONTHS
X $75 PER MONTH
= $900 PER YEAR
X 20 YEARS
$18,000 YEARS

$18,000 AT $75 PER MONTH
− 12,000 AT $50 PER MON
$6,000

TOMMY AND GINA CAN SAVE $6,000 MORE OVER 20 YEARS BY DEPOSITING $75 EACH MONTH RATHER THAN $50.

9. Describe two different ways to find the number of blank cells in the spreadsheet, one of which is multiplication.

	A	B	C	D	E	F	G	H
1								
2								
3								
4								
5								
6								

6 × 8 = 48 CELLS OR 8+8+8+8+8+8 = 48 CELLS

10. Did you multiply the number of rows by the number of columns or vice versa? Does it matter? What does that tell you? YES! - NO - THIS TELLS ME THAT YOU WILL END UP WITH THE SAME ANSWER NO MATTER THE ORDER IN WHICH YOU MULTIPLY

11. There are sixty fans with tickets to sit on this set of bleachers. How many fans that should sit in each row?

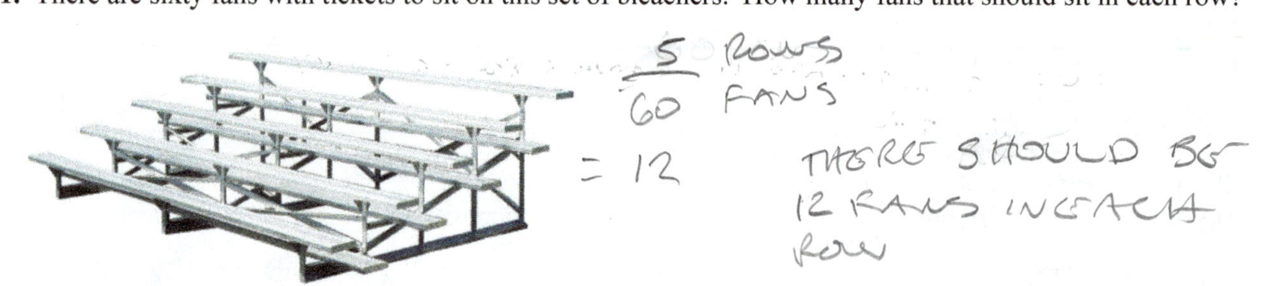

5 ROWS) 60 FANS = 12

THERE SHOULD BE 12 FANS IN EACH ROW

12. Does the order of your calculation matter in Question 11? What does that tell you?

NO IT TELLS ME THAT IT DOES NOT MATTER WHICH ORDER YOU DO THE MULTIPLICATION

13. List some words or phrases that are used to indicate that multiplication should be done.

14. List some words or phrases that are used to indicate that division should be done.

1-6 Portfolio Name ____________________

Check each box when you've completed the task. Remember that your instructor will want you to turn in the portfolio pages you create.

Technology

1. □ Create a spreadsheet that keeps track of the amount of money accumulated by Tommy and Gina (pg. 47). In column A, do it the long way: enter every monthly deposit over a 10-year period, as well as the $1,000 initial deposit. You don't have to enter each amount separately: use copy and paste. See a template to help you get started in the online resources for this lesson. At the top of columns C and E, create cells in which a user can enter the monthly deposit and the number of years. At the top of column F, enter a formula to calculate the total amount deposited (including the initial $1,000). A template to help you get started can be found in the online resources for this lesson.

Skills

1. □ Include any written work from the online skills assignment along with any notes or questions about this lesson's content.

Applications

1. □ Complete the applications problems.

Reflections

Type a short answer to each question.

1. □ Describe a situation outside of school where multiplying was useful.
2. □ Describe a plan you could follow to keep from getting too far behind in class work.
3. □ How can thinking about the cumulative effects of small things help you either in or outside of the classroom?
4. □ Take another look at your answer to Question 0 at the beginning of this lesson. Would you change your answer now that you've completed the lesson? How would you summarize the topic of this lesson now?
5. □ What questions do you have about this lesson?

Looking Ahead

1. □ Read the opening paragraph in Lesson 1-7 carefully, then answer Question 0 in preparation for that lesson.

1-6 Applications

Name ______________________

1. If you spend 30 minutes a day goofing off on the Internet during time you've set aside for studying, how much study time (in minutes, and in hours) will you waste over the course of a 16 week semester? What if you waste 45 minutes a day?

Gaining or losing weight comes down to calories burned vs. calories consumed. Burn more calories than you take in, and you'll lose weight. Burn less than you take in, and you'll gain weight. Simple. Let's study some aspects of weight change.

2. George weighed 170 lbs when he started college. If he gains just 0.25 lb each month for 4 years of college, how much will he weigh? What if he keeps that weight gain for the next 6 years after college?

3. A rule of thumb used by nutritionists is that to lose 1 lb of body fat, you need to burn 3,500 calories above what you take in. If you burn 400 more calories than you take in each day, how long will it take to lose 1 lb? What about 5 lbs?

4. An average-sized person will burn about 350 calories in an hour of walking at a fairly brisk pace. How many calories would you burn if you walk an hour a day for six months? How many pounds of body fat would that correspond to?

1-6 Applications Name WILLIAM BEZLER

5. The figure below is made up of some number of smaller squares, like this one: □

Calculate the number of small squares in the figure using two different methods. Write a brief description of each method.

I COUNT ~~THAT THERE IS~~ 4 ROWS OF SQUARES AND 5 COLLUMS

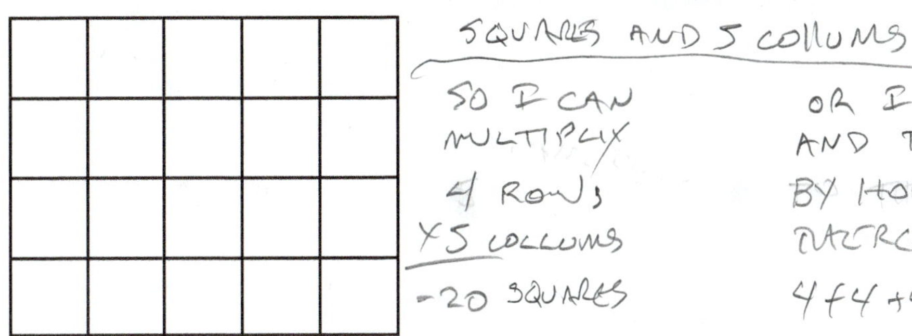

SO I CAN MULTIPLY 4 ROWS × 5 COLLUMS = 20 SQUARES

OR I CAN ADD THE 4 ROWS BY HOW MANY COLLOMS THERE ARE WICH IS 4 4+4+4+4+4 = 20

6. Now let's add an extra dimension – literally! The next figure is made up of some number of little cubes, like this one:

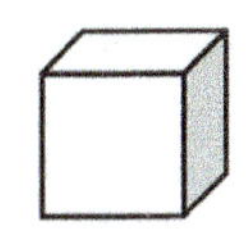

Calculate the number of small cubes in the figure using two different methods. Write a brief description of each method.

I SEE 3 ROWS OF SQUARES WITH 4 COLLUMS AND 2 SHEETS

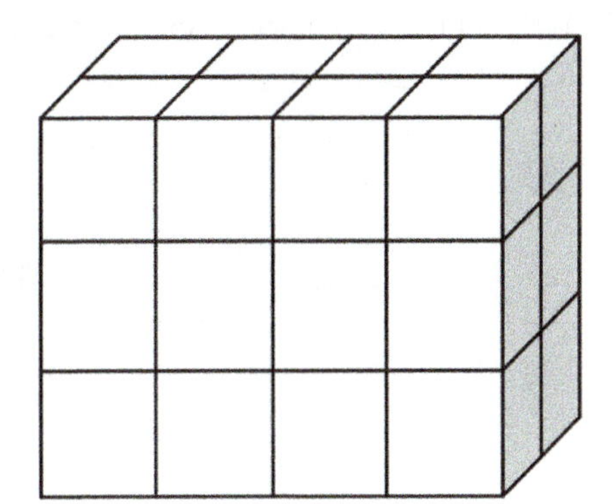

SO I CAN MULTIPLY 3 ROWS × 4 COLLUMS = 12 BOXES × 2 SHEETS = 24 SQUARES

OR I CAN ADD 3 ROWS BY HOW MANY COLLUMS THERE ARE WICH IS 4 3+3+3+3 = 12 AND THEN ADD 2 SHEETS DEEP 12+12 = 24 SQUARES

7. Look very closely at a TV screen (like from an inch away. And turn it on first!). What do you see? What does that have to do with the theme of this section?

A GIGANT AMOUNT OF TINY SQUARES IN ROWS AND COLLUMS

IT IS SIMULAR IN THE WAY THAT YOU COULD USE MATH FOR A EASIER METHOD TO FIND THE NUMBER OF SHAPES IN A GRID

Lesson 1-7 Avoiding Empty Pockets

Learning Objectives

- ☐ 1. Distinguish between simple interest and compound interest.
- ☐ 2. Interpret exponents as repeated multiplication.
- ☐ 3. Practice working with exponents and the order of operations.

The most powerful force in the universe is compound interest.
– Albert Einstein

Einstein was a smart guy that knew a little something about powerful forces, so that quote is a bit of an eye-opener. In this lesson, we'll study interest, a subject that everyone in our society needs to know something about (unless they want to end up like our friend to the right). If you have a trust fund that will afford you the opportunity to pay cash for a home, a car, a yacht ... good for you. Feel free to send us a generous donation, care of the publisher. If not, you'll need to borrow money many times during your lifetime, and the amount of interest you pay will be an important part of budgeting your money.

Then there's saving money, another important part of a financial plan. Putting money in your sock drawer is *not* a sound investment strategy: you'll want to invest in a variety of accounts that will add interest and grow your money. In either case, understanding the power of compound interest will set you on a path to financial success.

0. After reading the opening paragraph, what do you think the main topic of this section will be?

MONEY MANAGEMENT

1-7 Group

1. Discuss any experiences you've had with some of these important financial situations: being responsible for a credit card or checking account, borrowing money to buy a car or home, starting a savings or retirement account, or others. HAVING A SAVING ACCOUNT, CREDITCARD, AND LOANS

2. Each of the situations listed above is distinct, but they have some common traits. List any you can think of.

THE AMOUNTS OF EACH ALL CHANGE FROM INTEREST RATES

Interest is a fee paid for the use of someone else's money. If you borrow money to buy a car, you pay the lending institution for using their money at the time of purchase. If you put money into a savings account, the bank pays you for having access to your money, which they in turn can use to lend to other people. Interest is calculated as a percentage of the original amount of money. For example, if you deposit \$100 in an account that pays 4% interest per year, at the end of 1 year, you would have 4% of \$100 extra in the account. This is called **simple interest.**

Computing Simple Interest

To find the simple interest on an amount of money, you multiply the original amount by the interest rate (written as a decimal) and the amount of time. The units for the time (often years) should match the units on interest rate (often percent per year).

Compound interest, on the other hand (the force that Einstein was so enamored with), is interest that is paid not just on the original amount deposited, but on interest previously earned as well. To help you to understand the difference between simple and compound interest, let's look at two different accounts:

Account 1: You deposit $1,000 into an account that pays 5% simple interest on that $1,000 each year.

Account 2: You deposit $1,000 into an account that pays 5% of the current account balance in interest each year.

3. Answer the following questions about Account 1. Getting an answer is nice, but make sure that you think carefully about how the amount is growing and look for patterns.

How much interest will you earn in the...		How much will be in the account at the end of the...	
...first year?	$1,000 • .05 = $50	... first year?	1000 + 50 = $1,050
...second year?	$1,000 • .05 = 50 + 50 = $50	...second year?	$1,050 + 50 = $1,100
...third year?	1000 • .05 = 50 + 100 = $50	...third year?	$1,100 + 50 = $1,150

4. Repeat Question 3 for Account 2.

How much interest will you earn in the...		How much will be in the account at the end of the...	
...first year?	$1,000 • .05 = 50	... first year?	$1,050.
...second year?	$1,000 (1.05) = $52.50	...second year?	$1102.50 ~~1,052.50~~
...third year?	$1,102.50 • 1.05 $55.13	...third year?	$1,157.63

5. Fill in the next table by finding the amount in the account at the end of each number of years. You've already found four of the amounts in the table. Hopefully.

	Account 1	**Account 2**
Start	$1,000.00	$1,000.00
After 1 year	$1,050.00	$1,050.00
After 2 years	$1,100.00	$1,102.50
After 3 years	$1,150.00	$1,157.63
After 4 years	$1,200.00	$1,115.51
After 10 years	$1,500.00	$1,628.89

6. This is the calculation used to find the amount after the first year in either account:

$$\text{Amount} = \$1{,}000 + 0.05(\$1{,}000)$$

Can you think of a way to simplify the calculation so that all you have to do is one multiplication?

YES 1000 × .05 = 1,050.00

7. Why does the value of the compound interest account grow faster?

THE ACCOUNT WITH THE COMPOUND INTEREST GROWS FASTER, BECAUSE IT PAYS INTEREST ON THE BALANCE EACH YEAR, NOT THE ORIGINAL AMOUNT

8. Here's a way to describe the difference in growth of the two accounts: Account 1 increases in value by $50 every year, while Account 2 has its value multiplied by 1.05 every year.

So the value of Account 1 at any time looks like $1,000 + 50 × (the number of years passed), and the value of Account 2 at any time looks like $1,000(1.05)(1.05)..., where the multiplication occurs the same number of times as the number of years that have passed.

1-7 Class

1. Here's a key observation about the repeated multiplication that occurs when finding the value of Account 2 in the group activity: repeated multiplication can be expressed more concisely using an EXPONENT.

Think about an investment that earns compound interest at a rate of 12% per year. The value of a $10,000 investment in this account after 5 years can be found by evaluating the expression

$10{,}000(1.12)^5$

2. Without using a calculator, write that expression without any exponents.

10,000 X (1.12 · 1.12 · 1.12 · 1.12 · 1.12) =

1.12 X 1.12 · 1.12 X 1 · 1.2 X 1.12 =

3. What is the value of the $10,000 investment after 5 years?

$ 1,7623.42

4. If the interest paid is simple interest, the account value goes up by 0.12($10,000) = $1,200 per year. What would the value be after 5 years in that case?

$10000 + 1,2000 PER YEAR X 5

$1,200 PEAR YEAR
X 5 YEARS
= $6000
+10,000
= $16,000

A population of rabbits in an area can grow pretty quickly. In fact, such a population can double every four months. If a field starts with 50 rabbits, the population could double 3 times in a year. After 1 year, the population would be $50\cdot(2\cdot2\cdot2)$.

5. Write the expression using an exponent, then find the number of rabbits after 1 year.

$50\cdot(2)^3$ = 2 X 2 = 4 X 2 = 8 X 50 = 400

6. Given the numbers from Question 5, why is the Earth not overrun by rabbits?

THEY CAN STARVE TO DEATH, GET EATEN AS PREY OR DIE OF OLD AGE

7. If you borrow $500 at 3% simple interest, the amount you'd owe after 4 years is given by

$500 + 500(0.03)(4)$

To correctly evaluate that expression, you'll need to remember the order of operations, which says that

500.(00,3)(4) comes before + 500.

8. Evaluate the expression to find how much you'd owe.

.03
x 4
= .12
x 500
= 60 + $500 = $560

= $560.00

9. The standard order of operations is a set of rules agreed on for performing calculations with more than one operation. The point is to avoid having to use parentheses in every such situation. List the order of operations here:

PARENTHISIS
EXPONETS
MULTIPLY
DIVIDE
ADDITION
SUBTRACTIONS

10. Perform the calculation below, then compare your result to Question 8.

$500(1 + 0.03 \cdot 4)$

IT IS THE SAME

11. The reason you got the same answer for Questions 8 and 10 is the ASSOCIATIVE property of multiplication over addition. (You DID get the same answer, right?)

2x(3+6) = 18

OR 3+6 = 9² = 18

12. One of the things you were asked to estimate in Lesson 1-4 was the monthly payment on a $10,000 car if you wanted to pay it off in 3 years. If the interest rate is 3%, the expression below can be used to calculate the exact monthly payment.

$$\frac{\left(\frac{0.03}{12}\right)(10{,}000)\left(1+\frac{0.03}{12}\right)^{36}}{\left(1+\frac{0.03}{12}\right)^{36}-1}$$

Clearly, this is a pretty complicated expression that would require some help from a calculator or computer. But you can't get that help unless you understand the order of operations, because calculators and computers are programmed to obey those rules. Spend some time writing this expression the way you'd need to enter it into a calculator or spreadsheet. Use " ^ " for exponents, " * " for multiplication, and " / " for division.

13. Use the expression you wrote in Question 12 to find the payment. The correct answer is shown in the spreadsheet below. If you didn't get that answer, try to amend your expression in Question 12.

In the spreadsheet below, the payment is calculated in cell B4 by entering your formula from Question 12, with the following modifications: 36 gets replaced by B1, 0.03 with B2, and 10,000 with B3. You'll need this for the Technology part of your portfolio in this lesson.

	A	B
1	Months	36
2	Rate	0.03
3	Amount	$10,000.00
4	Payment	$290.81

IN EXCEL CLASS!

1-7 Portfolio

Name William Beeler

Check each box when you've completed the task. Remember that your instructor will want you to turn in the portfolio pages you create.

Technology

1. ☐ Create a spreadsheet based on the car payment calculations on page 58. You should be able to enter different values for the months, interest rate, and loan amount and calculate the payment automatically. A template to help you get started can be found in the online resources for this lesson.
2. ☐ Do a Web search for different ways to remember the order of operations and find one you like. Warning: you may come across some that are not exactly family-friendly.

Skills

1. ☐ Include any written work from the online skills assignment along with any notes or questions about this lesson's content.

Applications

1. ☐ Complete the applications problems.

Reflections

Type a short answer to each question.

1. ☐ Why do you think Einstein felt so strongly about compound interest?
2. ☐ What's the point of having an order of operations in math?
3. ☐ Take another look at your answer to Question 0 at the beginning of this lesson. Would you change your answer now that you've completed the lesson? How would you summarize the topic of this lesson now?
4. ☐ What questions do you have about this lesson?

Looking Ahead

1. ☐ Read the opening paragraph in Lesson 1-8 carefully, then answer Question 0 in preparation for that lesson.

1-7 Applications

Name WILLIAM BEEZLER

Due to a variety of factors including fuel prices and adverse weather patterns, the cost of many groceries is expected to increase by as much as 5% per year. Let's illustrate the effects on consumers using the example of a frozen pizza that currently costs $4.80.

1. If the price of the pizza does go up by 5% next year, find the amount of increase and the new cost.

$4.80 CURRENT PIZZA
X .05 % INCREASE
= .24¢ AMOUNT OF INCREASE

$4.80
= $5.04

AMOUNT OF INCREASE .24¢
NEW COST $5.04

Sentence?

2. Complete the table below by calculating the new price of several other items for the same 5% increase. Notice the formula in cell C4 which reminds you of the most efficient way to do this calculation.

C4 fx =B4*1.05

	A	B	C
1	Item	Old price	New price
2	Candy bar	$0.60	$0.63
3	Loaf of bread	$1.89	$1.98
4	Frozen pizza	$4.80	$5.04
5	Pound of sirloin	$6.99	$7.34
6	Olive Oil	$8.49	$8.89
7	Bag of chicken breast	$11.79	12.38

=B6 * 1.05
=B7 * 1.05

3. If the pizza goes up by 5% again the following year, fill in each blank to complete the calculation you'd use to find the price after two 5% increases.

$$4.80(1.\boxed{0}\boxed{5})^{\boxed{2}}$$

4. How much would the pizza cost after two 5% increases?

$4.80
X(1.05)²
$5.29

$5.29

1-7 Applications Name WILLIAM BEZZER

5. Fill in the table below, which shows the cost of the $4.80 pizza after each number of 5% increases. Thinking about your answer to Question 3 should help make the calculations go quicker.

# of increases	New price
1	$ 5.04
2	$ 5.29
3	$ 5.56
4	$ 5.83
10	$ 7.82

6. How does the price of the pizza after 10 years of 5% increases compare to what the price would be if it went up by just 5% of the original cost each year?

1	$4.80 X .05 = .24	$ 4.80 + .24 =	$5.04	$ 5.04
2		$ 5.04 + .24 =	$ 5.28	$ 5.29
3		$ 5.28 + .24 =	$ 5.52	$ 5.56
4		$ 5.52 + .24 =	$ 5.76	$ 5.83
10		$ 5.76 + .24 (6) =	$ 7.20	$ 7.82

IT WOULD BE .62 ¢ MORE

Bonus Question

Ice cream was traditionally sold in half-gallon containers, which contained 64 ounces of creamy deliciousness. Then around 2008, several major ice cream makers pulled a fast one on consumers: prices remained the same, but the size of the container was decreased to 56 ounces. Because the containers looked so similar, most people didn't notice that they were in effect paying more for ice cream. That went so well that containers were again decreased, to 48 ounces. Here's a challenging question: by what percent did the price effectively increase with each change in package size? (Hint: Make up a starting price for the 64 ounce package, then figure out how much 64 ounces would cost at the rate of that original price per 56 ounces.)

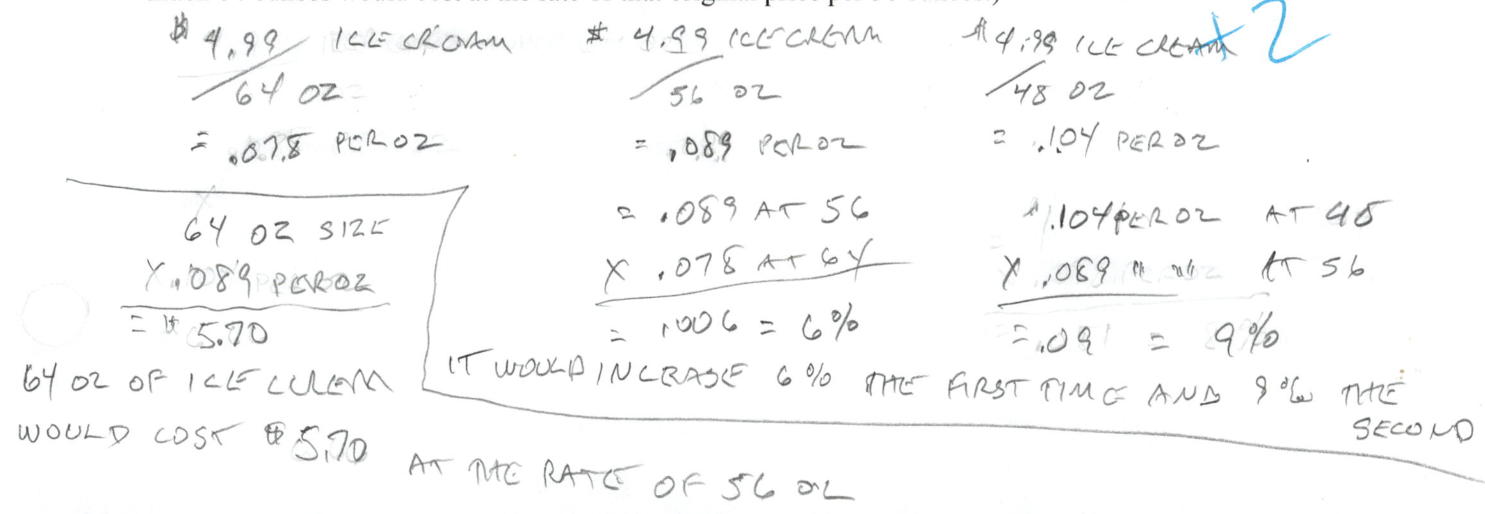

Lesson 1-8 Follow the Pattern

Learning Objectives

- ☐ 1. Recognize patterns and use them to make predictions.
- ☐ 2. Distinguish between linear and exponential growth.

"What we call chaos is just patterns we haven't recognized. What we call random is just patterns we can't decipher."
– Chuck Palahniuk

The world that we live in is a dynamic place: change is inevitable, and those that don't adjust and adapt tend to get left in the dust. In many walks of life, adapting and adjusting is based on recognizing patterns and trends. Societal trends, financial trends, behavioral trends ... if you can recognize these patterns, you have a much better chance at anticipating what's likely to happen next. This type of anticipation is often the edge you need against competitors, whether it's in school, a career, or your personal life.

I'm sure you know that people often ask in math classes, "When am I going to use this?" When they do, they're focusing too much on specific formulas and procedures, and not enough on the big picture. Being able to work through complex procedures and being able to carefully follow a series of instructions is important in almost any job, so we'll practice that in this lesson as well.

0. After reading the opening paragraph, what do you think the main topic of this section will be?

1-8 Group

How good are you at recognizing patterns? If you look for them, they're all around you. In some cases, there are patterns in things you see all the time, but never really thought about in terms of patterns. Two examples of visual patterns are shown here. The QR code will direct a smart phone to the website of the St. Louis Cardinals; the UPC is for a 12-pack of 7-Up. (Try them if you don't believe me.) They may look like a bunch of squares and lines at first glance, but there are definite patterns for the discerning eye that know what to look for.

One good way to practice recognizing patterns and anticipating where they lead is to study strings of numbers and letters, looking for a pattern. In Questions 1–8, find the pattern and use it to make a reasonable prediction for the next three numbers.

1. 3, 6, 9, 12, 15, 18, 21

2. 3, 6, 12, 24, 48, 96, 192

3. 320, 160, 80, 40, 20, 10, 5

4. 320, 240, 160, 80, 0, −80, −160

5. A, D, G, J, M, Q, T

6. 1, b, 3, d, 5, f, 7, H, 9

7. 1, 1, 2, 3, 5, 8, 13, 21, 34, 55

8. J, F, M, A, M, J, J, A, S

The familiar UPC codes that are used to find the price of an item have something called a **check digit** that's important in avoiding scanning errors. Obviously, it's important both to the store and the buyer that prices are entered correctly. To practice following a procedure, read the instructions below carefully, then use them to answer Questions 9 and 10.

UPCs are 12-digit codes. A 12-digit Universal Product Code (UPC) code has three parts. The UPC for a 12-pack of 7-Up was shown on the previous page: **0 78000 01180 7**.

- The first 6 digits of this code – 0 78000 – identify the manufacturer.
- The next 5 digits – 01180 – are the item number; identifying a specific product from this manufacturer.
- The 12th digit is the check digit. When the item is scanned, the reader performs a calculation on the first 11 digits, and if the result doesn't match the check digit, the machine knows something is wrong and the scan is rejected.

The calculation of the check digit is a 5-step process:

Calculating the Check Digit in a UPC

Step 1. Add all of the digits in the odd numbered positions – the 1st, 3rd, 5th, 7th, 9th, and 11th.

Step 2. Multiply the result from Step 1 by 3.

Step 3. Add the digits in the even numbered positions – the 2nd, 4th, 6th, 8th, and 10th. (Don't add the 12th digit!)

Step 4. Add the results of Steps 2 and 3.

Step 5. The check digit is the number you have to add to the result of Step 4 to get to the next multiple of 10 (that is, the next number that ends with a zero).

9. Check that the UPC for 7-Up provided above has the right check digit. Show all of your calculations!

0+8+0+0+1+0 = 9 = STEP 1 × STEP 2 9
7+0+0+1+8 = 16 = STEP 3
×3
27
+16
43 + 7 = 50

10. Repeat Question 9 for this UPC: **8 85909 59964 6.** (You'll be asked later to find what product the code corresponds to.)

8+5+0+5+9+4 = 31
8+9+9+9+6 = 41
31
×3
93
41
134 + 6 = 140

1-8 Class

1. If I offered you one of the two accounts detailed below, which would you choose, and why?

	Account 1	Account 2
Start	$1,000.00	$1,000.00
After 1 year	$1,060.00	$1,050.00
After 2 years	$1,120.00	$1,102.50
After 3 years	$1,180.00	$1,157.63
After 4 years	$1,240.00	$1,215.51

I WOULD CHOOSE ACOUNT 2 BECAUSE ACOUNT 2 IS COMPOUND INTEREST AND WOULD EVETUALY OUTGAIN ACOUNT 1 EVEN THOU IT IS A HIGHER INTEREST RATE

Let's take a more in-depth look at the patterns from the first two questions in this lesson.

2. Describe the change from one number to the next in this pattern: 3, 6, 9, 12, 15, 18, 21, ... Be *very* specific. That's way more than just specific. Way.

THE CHANGE FROM NUMBER TO THE NEXT IS JUST THE ADDITION OF 3 TO EACH NUMBER

3. Repeat Question 2 for this pattern: 3, 6, 12, 24, 48, 96, 192, ...

THE CHANGE FROM ONE NUMBER TO THE NEXT IS THAT THE NEXT NUMBER IS MULTIPLIED BY 2

In both of the lists above, the values get increasingly bigger, but in different ways.

4. In the first list, the values grow by ADDING the same constant number.

This type of growth is called **linear growth.**

5. In the second list, the values grow by MULTIPLYING by the same constant number.

This type of growth is called **exponential growth.**

The spreadsheet to the right shows an example of linear growth.

	A
1	35
2	45
3	55
4	65
5	75
6	

6. What's the next value on the list? 

7. Write a verbal description of how you would obtain the value in cell A5 from the value in cell A5.

A5+10=B6

ADD A5+10=75

(65)

8. What would you type in cell A6 to calculate this value?

=A5+10

The spreadsheet to the right shows an example of exponential growth.

	A
1	2
2	20
3	200
4	2,000
5	
6	

9. What's the next value on the list?

\# 20,000

10. Write a verbal description of how you would obtain the value in cell A6 from the value in cell A5.

MULTIPLY 2,000 X 10 = 20,000

11. What would you type in cell A6 to calculate this value?

=A4 X 10 = A6 = 20,000

1-8 Portfolio

Name ______________________________

Check each box when you've completed the task. Remember that your instructor will want you to turn in the portfolio pages you create.

Technology

1. ☐ Five scenarios are illustrated by a portion of a spreadsheet in the Applications section of this lesson. For the first four, create a spreadsheet that will calculate the next ten entries on the list. A template to help you get started can be found in the online resources for this lesson.
2. ☐ Do an Internet search to find the product that has the UPC provided in Question 10 on page 64.

Skills

1. ☐ Include any written work from the online skills assignment along with any notes or questions about this lesson's content.

Applications

1. ☐ Complete the applications problems.

Reflections

Type a short answer to each question.

1. ☐ Explain the difference between linear growth and exponential growth in your own words.
2. ☐ How do linear and exponential growth apply to what we learned about simple and compound interest in Lesson 1-7?
3. ☐ Describe one or two situations outside of school where recognizing a pattern would be useful. Be specific!
4. ☐ Take another look at your answer to Question 0 at the beginning of this lesson. Would you change your answer now that you've completed the lesson? How would you summarize the topic of this lesson now?
5. ☐ What questions do you have about this lesson?

Looking Ahead

1. ☐ The terms below will be discussed in Lesson 1-9 in terms of survival skills for college and beyond. If you're unfamiliar with any of these terms, look them up on the Web in preparation for the next class.

Critical thinking	Entrepreneurialism
Problem solving	Written communication
Collaboration	Oral communication
Network	Curiosity
Agility	Imagination
Adaptability	Initiative

2. ☐ Read the opening paragraph in Lesson 1-9 carefully, then answer Question 0 in preparation for that lesson.

1-8 Applications

Name ______________________

Identify each scenario as illustrating either linear growth, exponential growth, or neither. Show enough work to justify your answer, and if you can, find the next value on the list.

1.

	A	B
1		Taxi Fare
2	Start	$3.30
3	After 1 mile	$5.70
4	After 2 miles	$8.10
5	After 3 miles	$10.50
6	After 4 miles	$12.90
7	After 5 miles	$15.30

2.

	A	B
1		Account value
2	Start	$12,000.00
3	After 1 year	$13,200.00
4	After 2 years	$14,520.00
5	After 3 years	$15,972.00
6	After 4 years	$17,569.20
7	After 5 years	$19,326.12

3.

	A	B
1		Population
2	Start	10,000
3	After 1 year	10,200
4	After 2 years	10,404
5	After 3 years	10,612
6	After 4 years	10,824
7	After 5 years	11,041

4.

	A	B
1		Salary
2	Start	$40,000
3	After 1 year	$41,500
4	After 2 years	$43,000
5	After 3 years	$44,500
6	After 4 years	$46,000
7	After 5 years	$47,500

5.

	A	B
1		Stock price
2	Start	$40.00
3	After 1 month	$42.00
4	After 2 months	$44.50
5	After 3 months	$48.00
6	After 4 months	$52.00
7	After 5 months	$60.00

Lesson 1-9 Survival Skills

Learning Objectives

- ☐ 1. Identify important skills for college students to have.
- ☐ 2. Understand units for area and volume.
- ☐ 3. Convert units by multiplying and dividing.

Even if you're on the right track, you'll get run over if you just sit there.
– Will Rogers

If you're on your own in the wild, survival is about finding food, water and shelter, and avoiding perilous situations. In the wild world of academia, the challenges aren't typically life-threatening, but that doesn't mean you can just wander around without survival skills and a plan. In this lesson, we'll focus on some skills that are necessary for success in modern higher education. We'll then practice some of those skills in the context of numbers and geometry. As always, try to focus on the skills rather than specific procedures or formulas. Your (academic) survival just might depend on it.

0. After reading the opening paragraph, what do you think the main topic of this section will be?

1-9 Class

In his 2010 book *The Global Achievement Gap,* educational consultant Tony Wagner outlines seven survival skills that students have to master to reach their academic potential. Define what each of these skills means to you.

1. Critical thinking and problem solving

CRITICAL THINKING IS SELF BLAME/NON PRODUCTIVE

PROBLEM SOLVING IS WORKING TOWARDS A RESALUTION

2. Collaboration across networks and leading by influence

COLLABORATION IS WORKING IN A TEAM

LEADING BY INFLUENCE

3. Agility and adaptability

AGILITY IS BEING FLEXABLE AND QUICK

ADAPTABILITY IS BEING ABLE TO FIT IN OR MAKE OTHER METHODS WORK

4. Initiative and entrepreneurialism

INITIATIVE IS TAKING CHARGE

ENTRE..... IS CAPITULIZEING ON T

5. Accessing and analyzing information

ACC = LOOKING UP OR

ANAL = EXAMANING INTO

6. Effective written and oral communication

BOTH = MAKING THAT UNDERSTANDABLE

7. Curiosity and imagination

CUR = SEEKING OR TESTING

IMAG = CREATING NEW THOUGHTS

Here's a key quote from Wagner's book: *"Today knowledge is ubiquitous, constantly changing, growing exponentially … . Today knowledge is free. It's like air, it's like water. It's become a commodity … . There's no competitive advantage today in knowing more than the person next to you. The world doesn't care what you know. What the world cares about is what you can do with what you know."*

8. What does this quote mean to you? Do you agree with it? Why or why not?

NOTHING, ITS PHYLOSAFEE – NO I DONT – I CANT SEE IT
OR PROVE IT.

1-9 Group

This activity is a bit less guided than most: the idea here is to promote some of the qualities Dr. Wagner wrote about, like problem solving, agility, adaptability, initiative, imagination, and (of course) collaboration, which is always one of the key goals in group work.

We'll begin by working with area and volume. **Area** is the measure of size for two-dimensional objects, like floors, walls, posters, fields, etc. **Volume** is the measure of size for three-dimensional objects, like beer mugs, swimming pools, buildings and so on.

1. Write out the calculation of the area of this figure in two different ways: first using addition, then using multiplication.

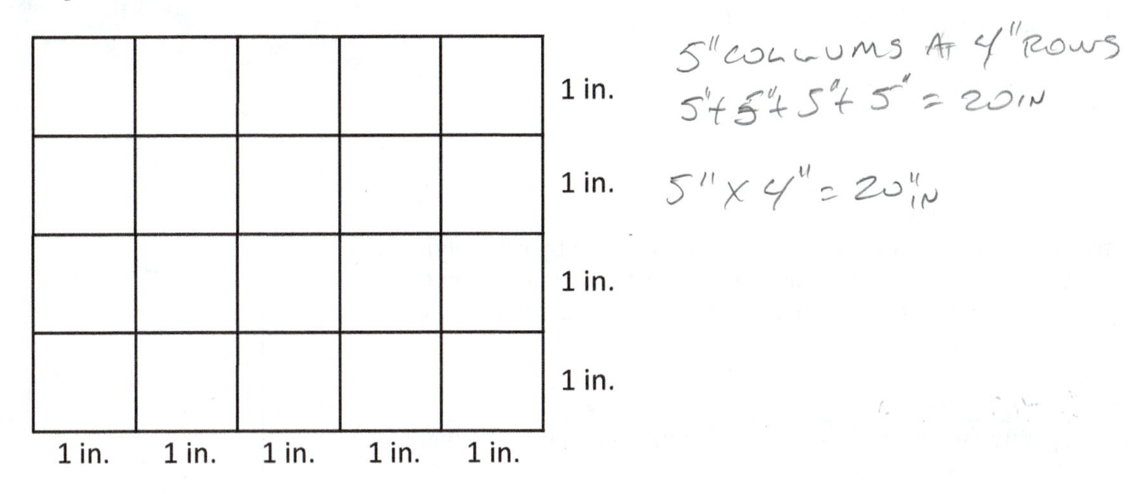

2. When length is measured in inches, the correct units for area are ____________ ____________.

3. Write out the calculation of the volume of this figure in two different ways: first using addition, then using multiplication.

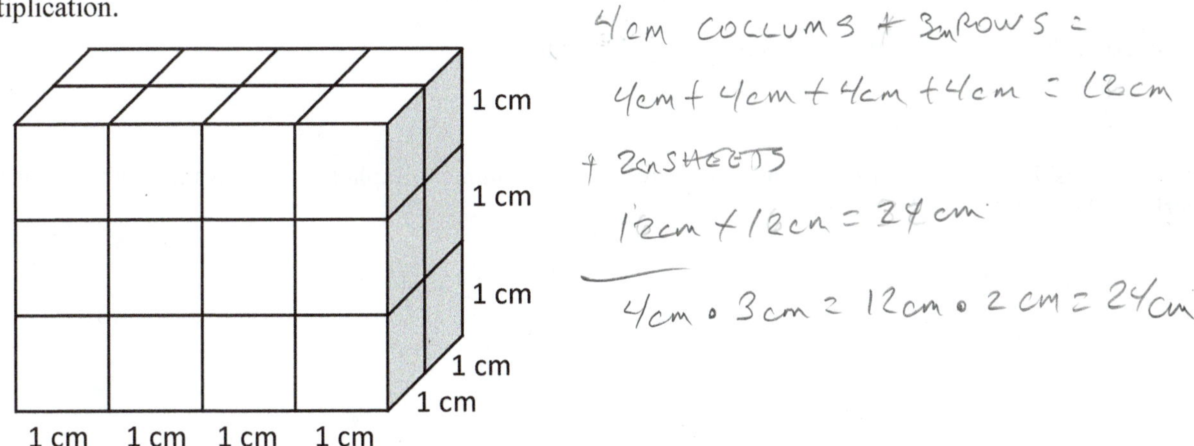

4. When length is measured in centimeters, the correct units for volume are ____________ ____________.

5. My friend David Najar is an acclaimed artist. The value of a painting is to some extent determined by the area of the canvas. A Najar piece I recently acquired has dimensions as shown, and you can find the area by multiplying the length and width. What is the area?

Area = 24 × 32 (Include units on each number.)

Area = 768 u²

The area units for the canvas are SQUARE INCHES

because there are 2 factors multiplied together, each of

which has INCHES as units.

24 in.

32 in.

6. The volume of a cylindrical oil storage tank can be found by multiplying π by the square of the radius times the height. Use the second diagram (not one of my best, but to be fair I'm a mathematician, not a graphic artist) to find the volume of the tank.

Volume = 3.1417 × 5 × 5 × 9
(Write four separate factors, and again include units on all measurements.)

Volume = 706.9 cm
(Round to the nearest tenth.)

The volume units for the oil tank are METERS CUBED

because there are 3 factors that have METERS as units.

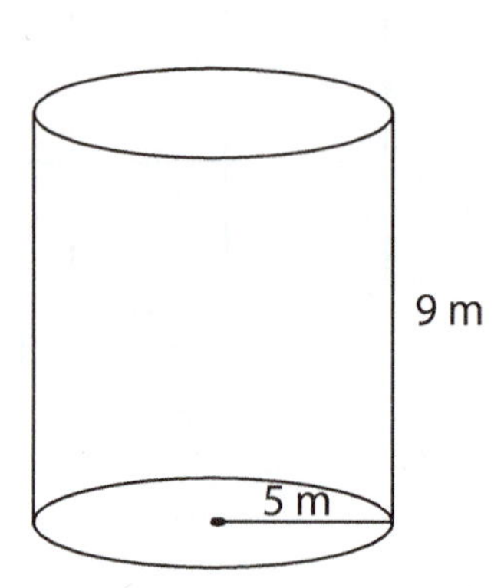

9 m

In Questions 7–12, decide if the calculation is for area or volume. Explain your reasoning. You don't have to perform the calculation.

7. $(4\text{m})(6\text{m})$ THERE ARE 2 FACTORS = AREA

8. $(3\text{ft})(5\text{ft})(10\text{ft})$ THERE ARE 3 FACTORS = VOLUME

9. $\dfrac{\pi(4\text{ in.})^3}{3}$ (Think about how the exponent affects the units!)

= VOLUME

THERE ARE 3 FACTORS

10. $2\left(\frac{4}{3}(1.5\text{m})^3\right)+\pi(1.5\text{m})^2(3\text{m})$

THERE ARE MORE THAN 2 FACTORS

= VOLUME

11. $2\pi(1.5\text{m})^2 + 2\pi(1.5\text{m})(3\text{m})$

12. $\frac{1}{2}(10\text{ft})(18\text{ft}) + (10\text{ft})^2$

An understanding of the significance of units is useful in a number of different ways. It can help you to decide what calculation might be appropriate in a situation, and can also provide clues as to whether you're doing a calculation correctly. Also, if you're trying to find something like a distance, without units an answer makes no sense. If you ask me how far away I live from work and I answer "about four," that answer would tell you exactly nothing. Four what? Miles? Minutes? Streets? Houses?

Another important aspect of working with units is converting between different units. For example, you know that 1 minute and 60 seconds is the same amount of time. (We say that 1 minute is **equivalent** to 60 seconds.) So you could convert a time of 5 minutes into seconds by multiplying 5 by 60. And you could convert a time of 300 seconds into minutes by dividing 300 by 60. The next page lists a variety of conversion equivalences that we'll need from time to time. Use it to perform each unit conversion in Questions 13–20. Make sure you write out the calculation that you're doing, not just an answer. In some cases, you'll have to do more than one multiplication or division.

13. 300 sec = ____________ min

14. 3 gal = ____________ quarts

15. 54 m = ____________ yds

16. 215,424 in. = ____________ mi

17. 3.4 hrs = ____________ min

18. 5'2" = ____________ in.

19. 95 kg = 3,344 oz

20. 1,100 mm = 3,608 ft

$\frac{95\text{ kg}}{1} \cdot \frac{2.2\text{ lb}}{1\text{ kg}} \cdot \frac{209\text{ lb}}{1} \cdot \frac{16\text{ oz}}{1\text{ lb}}$

$\frac{3344\text{ oz}}{1} = 3,344\text{ oz}$

$\frac{1100\text{ mm}}{1} \cdot \frac{1\text{ m}}{1000\text{ mm}} \cdot \frac{1100\text{ m}}{1,000} \quad \frac{1.1\text{ m}}{1}$

$\frac{1.1\text{ m}}{1} \cdot \frac{3.28\text{ ft}}{1\text{ m}} = \frac{3.608\text{ ft}}{1} = 3,608\text{ ft}$

English Measure and Equivalents

Length	Weight
12 inches (in) = 1 foot (ft) 3 feet = 1 yard (yd) 5280 feet = 1 mile (mi)	16 ounces (oz) = 1 pound (lb) 2000 pounds = 1 ton (T)
Liquid Volume	**Time**
3 teaspoons (tsp) = 1 tablespoon (tbs) 8 fluid ounces (oz) = 1 cup 2 cups (c) = 1 pint (pt) 2 pints = 1 quart (qt) 4 quarts = 1 gallon (gal)	60 seconds (sec) = 1 minute (min) 60 minutes = 1 hour (hr) 24 hours = 1 day 7 days = 1 week 52 weeks = 1 year 365 days = 1 year

The Metric System

Kilo (1,000)	Hecto (100)	Deka (10)	Base Unit	Deci (1/10)	Centi (1/100)	Milli (1/1,000)
km	hm	dam	**Length** Meter (m)	dm	cm	mm
kg	hg	dag	**Weight** Gram (g)	dg	cg	mg
kL	hL	daL	**Volume** Liter (L)	dL	cL	mL

Conversions Between Systems

Length	Weight	Volume
2.54 cm = 1 in 1 m = 3.28 ft 1.61 km = 1 mi	28.3 g = 1 oz 2.2 lb = 1 kg	1.06 qt = 1 L 3.79 L = 1 gal

Math Note

In Unit 2, we'll study an important method of converting units called **dimensional analysis;** it can be used for tons of useful things. Tons!

1-9 Portfolio

Name ____________________

Check each box when you've completed the task. Remember that your instructor will want you to turn in the portfolio pages you create.

Technology

1. ☐ Create a spreadsheet that converts feet to meters, pounds to kilograms, and gallons to liters. Use the equivalences on page 74. A template to help you get started can be found in the online resources for this lesson.
2. ☐ There are many websites that will do unit conversions. Do an Internet search to find a conversion site, and use it to check your answers to Questions 13–20. Copy and paste the URL of the site you used to the spreadsheet document created in Question 1 so your instructor can see what you did.

Skills

1. ☐ Include any written work from the online skills assignment along with any notes or questions about this lesson's content.

Applications

1. ☐ Complete the applications problems.

Reflections

Type a short answer to each question.

1. ☐ Pick two of the seven survival skills from the beginning of the section that you think are most important to you. Discuss why you chose them, and how strong you think you are in these skills.
2. ☐ Why are units so important in measurement? Giving some examples of measurements with no units should help.
3. ☐ What does it mean to say that two measurements are equivalent? Why are equivalent units so useful?
4. ☐ Take another look at your answer to Question 0 at the beginning of this lesson. Would you change your answer now that you've completed the lesson? How would you summarize the topic of this lesson now?
5. ☐ What questions do you have about this lesson?

Looking Ahead

1. ☐ Read the opening paragraph in Lesson 1-10 carefully, then answer Question 0 in preparation for that lesson.

9/10

1-9 Applications

Name WILLIAM BEEZER

Tennis balls are packaged in a cylindrical can, and since they're round, that leaves a lot of empty space in the can. How much space? The radius of a tennis ball is 3.3 cm; the can has the same radius for a nice tight fit, and the height of the can is six times that radius, which is 19.8 cm.

1. The calculation below is intended to find the volume of the empty space in the can, but there's something wrong with it. Explain how you can tell without doing any calculations or looking up any formulas that the answer will be wrong. (Hint: The terms "units" and "like terms" might come into play.)

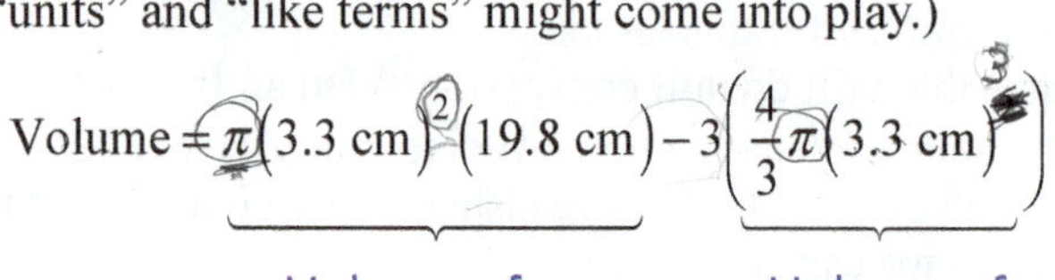

I CAN TELL WITHOUT DOING ANY CALCULATIONS, BECAUSE THE SECOND EQUATION SHOW THE UNITS SQUARED AND NOT WRITTEN IN UNITS CUBED

(1)

2. Find the fix needed to correct the mistake in the calculation. If you need help, do a Web search for "volume of a sphere." Then find the volume of empty space in the can. Round to the nearest hundredth.

$\pi(3.3\text{cm})^2(19.8\text{cm}) - 3\left(\frac{4}{3}\pi(3.3\text{cm})^3\right)$

$3.3^2\text{cm} \times 3.14 = 34.19\text{ cm}^2$

$34.19\text{ cm}^2 \times 19.8\text{ cm} = 677.05\text{cm}^3$

$3.3^3\text{cm} \times 3.14 = 112.84\text{ cm}^3$

$112.84\text{cm}^3 \times \frac{4}{3} = 150.46\text{ cm}^3$

$150.46\text{ cm}^3 \times 3 = 451.37\text{ cm}^3$

677.05 cm^3 = VOL OF CAN
-451.37 cm^3 = VOL 3 BALLS
$= 225.68\text{ cm}^3$ = EMPTY SPACE

= 225.8 WITH π BUTTON CALCULATOR

LAURA SAYS TO FORGIVE ME!

I DID NOT KNOW HOW TO USE THE PIE BUTTON

3. Use the unit converter website you found in the Technology portion of the portfolio page to convert this volume from cubic centimeters to liters.

225.68 cm^3 = .22568 LITERS

$225.68\text{cm}^3 \times .001$ = .22568

4. What percentage of the volume of the can is filled by the tennis balls? The percentage is the volume of the 3 tennis balls divided by the volume of the can, written in percent form. When you set up the division, include units on each volume. Round to the nearest hundredth.

$\frac{451.37\text{ cm}^3 \text{ VOL 3 BALLS}}{667.05\text{cm}^3 \text{ VOL CAN}}$

= .6767 = 67.67%

THE PERCENTAGE OF THE VOLUME OF THE CAN FILLED WITH TENNIS BALLS IS 67.67%

5. After performing the division in Question 4, what units remain? Do you think this will always happen when finding percentages?

~~THE UNIT THAT REMAINS IS THE PERCENTAGE OF THE EMPTY SPACE IN THE CAN~~ YES?

No units remain because they cancel out.

1-9 **Applications** Name WILLIAM BERGER

6. Racquetballs have a radius of about 2.8 cm. If three balls are packaged in a can like we described for tennis balls earlier, find each volume. Round to the nearest hundredth.

a. The volume of one ball.

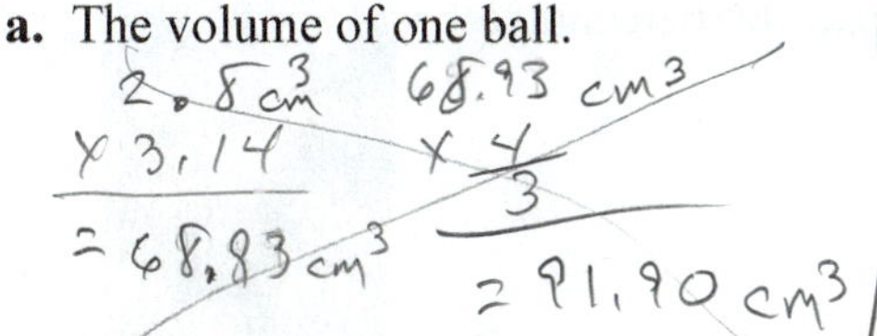

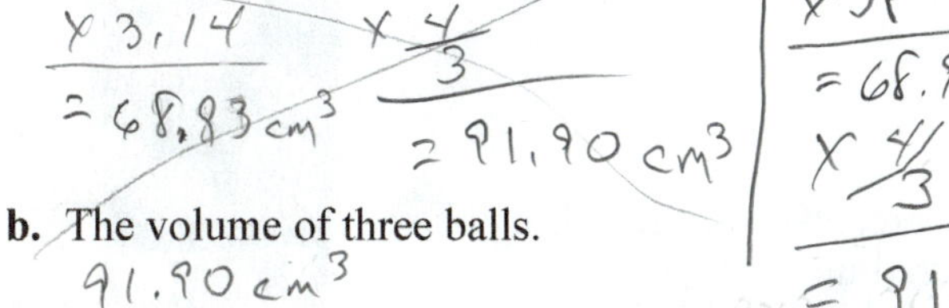

$2.8^3 \times \pi = 68.96\ cm^3 \times \frac{4}{3} = 91.95\ cm^3$

b. The volume of three balls.

$91.90\ cm^3 \times 3 = 275.72\ cm^3$

c. The volume of the can. (Remember, the height is six times the radius.)

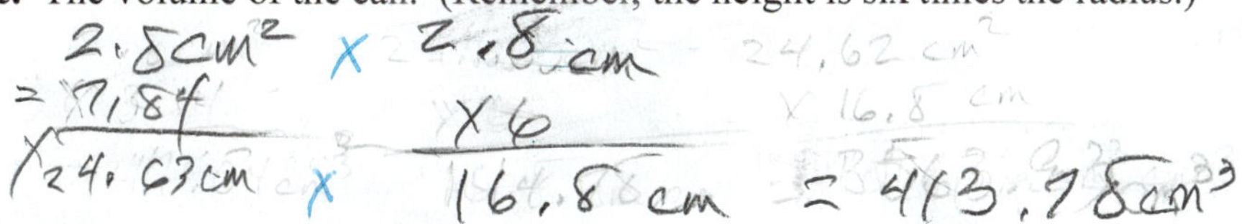

$= 413.78\ cm^3$

d. The volume of empty space in the can.

413.783 cm³ VOL CAN
− 275.72 cm³ VOL 3 BALLS
138.06 cm³ = VOLUME OF EMPTY SPACE IN CAN

e. The percentage of the can filled by the racquetballs. How does it compare to the tennis ball percentage? Does this surprise you?

$\frac{275.72\ cm^3 \text{ VOL 3 BALL}}{413.62\ cm^3 \text{ VOL OF CAN}}$ RACQUETBALLS

= .6666 = 66.6%

67.67% = TENNIS BALL

THE TENNIS BALLS TAKE UP APPROXIMENTLY 1% MORE THAN THE RACQUET BALL

NO, IT DOES NOT SURPRISE ME

Bonus Question: Golf balls also come in packages of three, but they are usually packaged in a rectangular box with a square bottom. The radius of each ball is about 2.2 cm. The length and width of the box's square base are twice that radius, and the height is six times the radius. You can find the volume of a rectangular box by multiplying length, width and height.

a. Would you guess that there would be more or less empty space by percentage than there is in a tennis ball can? Why?

b. Use your procedure from Question 4 to find the percentage of the golf ball box that is filled. Round volumes to the nearest hundredth.

Average thickness of a regular piece of paper: 0.05 millimeters

2. With some practice, scientific notation is a more intuitive way to represent all the numbers above. The notation involves powers of ten. To begin, find the pattern in the table below and put the appropriate exponents in the blanks. Check your results on a calculator.

1,000,000	$\times 10^6$
100,000	$\times 10^5$
10,000	4
1,000	3
100	2
10	1
1	0
0.1	-1
0.01	-2
0.001	-3
0.0001	-4
0.00001	-5

3. a) For numbers greater than one the power of ten is POSITIVE.

   b) For numbers less than one the power of ten is NEGATIVE.

   c) What power of ten gives an answer of one? 0

Scientific notation takes advantage of these relationships using the following:

In *scientific notation*, write numbers in the following form:

where N is a number between 1 (included) and 10 (not included) and n is an integer.

Use the following information to answer questions 4 and 5.

- The weight of the Earth is 13,200,000,000,000,000,000,000,000 pounds.
- The length of a hydrogen atom is 0.000000031 millimeters.

Scientific Notation Supplement: Large and Small Numbers

Scientific notation is used to write numbers that are very large or very small. This notation makes numbers that are difficult to work with, easier to work with, thereby making calculations easier. For example, in the sciences there are often very large or very small numbers. In chemistry, Avogadro's Number, which is 6022 with 20 zeros after it, is much easier to write using scientific notation. Without scientific notation, it can be difficult to perform calculations with this very large number. In this section we will learn how to use scientific notation to allow us to represent large and small numbers and do calculations with these numbers.

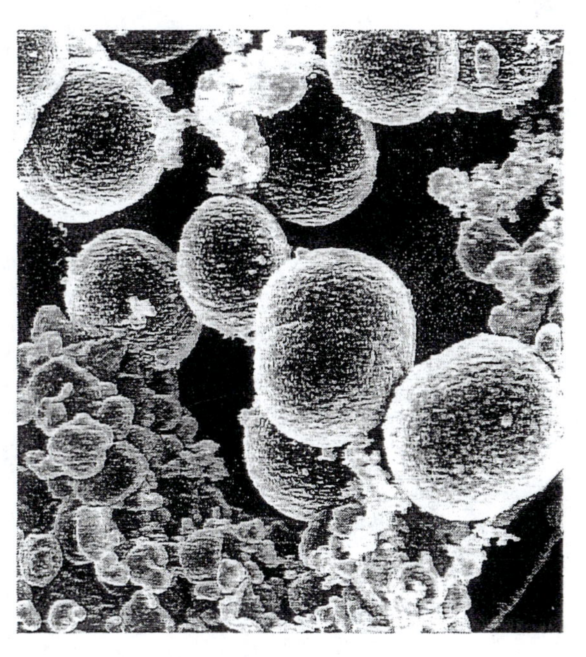

Public Health Image Library

As an additional resource visit the website Khan Academy and search for **Introduction to scientific notation** for a great video and description of scientific notation.

Scientific Notation: Class

1. At the bottom of the page are some common examples of very large and very small numbers written in ***standard form***, that is not written in scientific notation. Try adding, subtracting, multiplying or dividing some of these numbers with or without a calculator. Discuss any problems that your group finds with performing these computations when the numbers are in standard form.

DISTANCE TO PLUTO X DIAMETER GRAIN SAND

3,670,000,000 miles

X .0024 INCHES

= 8,808,000

Large Numbers

Average distance to Pluto: 3,670,000,000 miles

Estimate of cells in the human body from Smithsonian Magazine: 37,200,000,000,000

Distance of the sun from the nearest star: 39,900,000,000,000 kilometers

Avogadro's Number: 602,200,000,000,000,000,000,000

Small Numbers

Mass of a dust particle: 0.000000000753 kilograms

Diameter of a grain of sand: 0.0024 inches

BETWEEN 1 AND 10

Use the following information to answer questions 4 and 5.

- The weight of the Earth is 13,200,000,000,000,000,000,000,000 pounds.
- The length of a hydrogen atom is 0.000000031 millimeters.

4. The weight of the Earth written in scientific notation would be 1.32×10^{25}.

5. The length of a hydrogen atom in scientific notation would be 3.1×10^{-8}.

6. Convert the numbers mentioned at the start of the lesson into scientific notation.

   a) Average distance to Pluto: 3,670,000,000 miles

   3.67×10^{9}

   b) Estimate of cells in the human body from Smithsonian Magazine: 37,200,000,000,000

   3.72×10^{13}

   c) Distance of the sun from the nearest star: 39,900,000,000,000 kilometers

   3.99×10^{13}

   d) Avogadro's Number: 602,200,000,000,000,000,000,000

   6.022×10^{23}

   e) Mass of a dust particle: 0.00000000753 kilograms

   7.53×10^{-10}

   f) Diameter of a grain of sand: 0.0024 inches

Scientific Notation: Group Work

7. Convert the following numbers to scientific notation:

   a) 327,000 = 3.27×10^{5}

   b) 98,251,000 = 9.8251×10^{7}

   c) 0.00083 = 8.3×10^{-4}

   d) 0.00000125 = 1.25×10^{-6}

   e) 4,300,000,000 = 4.3×10^{8}

   f) 0.000000000008 = $8. \times 10^{-12}$

Scientific Notation: Class Revisited

Using calculators to compute products and quotients of numbers in scientific notation
Note: The textbook has directions for scientific notation on page 289.

> On your calculator, look for a button that looks like the following: (page 289 in text)
> EE
> These buttons represent "times 10" in your calculator.
>
> If you want to enter the number 2.3×10^{12} into your calculator, you type in:
> 2.3 [EE] 12 [=]

Look closely at how your calculator represents the number on its screen. Remember how this looks for when you perform calculations and get this notation in the future.

3.2 [X10^N] [PWR] OVER 2 [XN] [ENT]

Perform the following calculations with your calculator. Write your results in scientific notation.

12. $(3.2\times10^{13})(2\times10^{7})$ = ← Use these calculator commands: 3.2 [EE] 13 [X] 2 [EE] 7 [=]

6.4×10^{20}

13. $(3.2\times10^{13})\div(2\times10^{7})$ = ← If the calculator can fit an answer on its screen without writing it in scientific notation, it will do so. You will have to convert it back to scientific notation in this case.

3.2 [XN] OVER ÷ (2×10^{7}) =

16×10^{5}

14. $(6.45\times10^{15})\times(2.8\times10^{12})$ =

1.806×10^{28}

15. $(4.21\times10^{17})\div(8.42\times10^{10})$ =

5×10^{6}

16. $(4.7\times10^{-15})\times(5\times10^{-6})$ =

2.35×10^{-20}

WILLIAM BERGER

Scientific Notation: Applications

1. Convert the following numbers to scientific notation:

a) 2,543,000 = 2.543×10^6

b) 371,400 = 3.714×10^5

c) 0.0075 = 7.5×10^{-3}

9

d) 0.0000391 = 3.91×10^{-5}

e) 290,000,000 = 2.9×10^8

f) 0.00000000055 = 5.5×10^{-10}

2. Convert the following numbers to standard form:

a) 3.91×10^5 = 391,000

b) 2.593×10^9 = 2,593,000,000

c) 2.15×10^{-4} = .000215

d) 7.6×10^{-8} = .000000076

3. Write the following quantities without an exponent: STANDARD FORM

a) In 2015, Facebook reported that 3.5×10^8 photos were uploaded each day.

350,000,000

b) The website Tinder is estimated to be worth $\$1.6 \times 10^9$ by JMP Securities. (BusinessInsider)

$1,600,000,000

c) The average snail can move 1.81×10^{-3} miles in 5 hours.

.00181 ~~X 10^-3 = .00905 MILES IN 5~~ HOURS

4. The mass of a dust particle is about 7.53×10^{-10} kg. The mass of an electron is about 9.11×10^{-31} kg. Which object has a smaller mass?

$7.53 \times 10^{-10} > 9.11 \times 10^{-31}$

AN ELECTRON IS A SMALLER MASS BECAUSE IT HAS A LARGER NEGATIVE EXPONENT

Lesson 1-10 Did You Pass the Test?

Learning Objectives

- ☐ 1. Consider strategies for preparing for and taking math tests.
- ☐ 2. Understand the impact of a single question, or a single exam.
- ☐ 3. Calculate and interpret measures of average.

Tell me and I'll forget; show me and I may remember; involve me and I'll understand.
– Chinese proverb

When's the last time you were tested? If you think of that only in an academic sense, you may be consulting a calendar. If you're more of a metaphorical person, you may be thinking "I get tested every day." Life tests us in many different ways, and how we deal with tests goes a long way to determining our character. In fact, that's one of the reasons that an education is so important. The tests that you take in college classes are a metaphor for some of the most important events in life: with the pressure on, can you perform your best? What have you learned, and can you apply that knowledge when you need to?

0. After reading the opening paragraph, what do you think the main topic of this section will be?

1-10 Class

In this lesson, we're going to look at exams in a math course. How do you prepare? How is your grade for the exam determined? What impact does this score have on your grade in the course? How does an instructor determine how well the class as a whole understood the material?

Answer each question in your groups and then prepare to share your answers with the class.

1. What are some strategies you use to prepare for an exam?

2. What are some habits you have or have seen that are not helpful when preparing for an exam?

3. What are some strategies you use during an exam?

1-10 Group

The spreadsheet summarizes the results for one student on a 15-question math test with partial credit awarded.

1. Complete the totals at the bottom of each column using addition.

2. What percent of the total possible points did this student earn? (Round all percentages in this lesson to one decimal place.)

3. What letter grade did this student get? (The grading scale is 90% = A, 80% = B, 70% = C, 60% = D.)

4. Problem 15 on the test was an application (what some folks call "story" or "word" problems). What letter grade would this student have earned if she'd wimped out and skipped that question?

	A	B	C
1	Problem Number	Points Possible	Points Earned
2	1	2	2
3	2	2	2
4	3	2	0
5	4	5	4
6	5	5	2
7	6	8	4
8	7	8	3
9	8	9	9
10	9	4	4
11	10	5	5
12	11	5	5
13	12	6	6
14	13	6	6
15	14	10	6
16	15	10	10
17	Totals		
18			
19			

5. What's the highest grade you could get on this test if you skip problem 15 because it's a scary word problem?

6. Does Question 5 sound familiar to you? Have you ever been in that position? Explain.

7. What do these questions make you think about in terms of taking tests, and the importance of trying every question?

8. When a test has been graded and returned, students (and instructors) are usually interested in knowing what the "class average" was. Explain the meaning of the word "average" in your own words. Don't think in terms of formulas, but what it tells you.

	A	B	C
1	**Student**	**Exam 1 (%)**	**Exam 2 (%)**
2	Michael	80	89
3	Andy	77	93
4	Pam	68	84
5	Jim	81	88
6	Dwight	96	91
7	Stanley	54	75
8	Phyllis	75	54
9	Kevin	81	86
10	Creed	71	0
11	Darryl	89	83
12	Gabe	56	64
13	Toby	81	65
14	Holly	92	73

In math and stats, the term "average" is sort of a generic term, with several interpretations. Loosely defined, the average of a list of numbers is the most typical value. But what exactly does that mean? The one that appears most often? The one right in the middle? Or some mathematical combination? Use the test grades in the spreadsheet to answer Questions 9–11.

The **mean** of a set of numbers is what you probably think of as the "average." You find it by adding all of the numbers, then dividing by how many numbers there are on the list.

9. Find the mean test score on Exam 1.

The **median** of a list of numbers is the value that lives right in the middle of the set if it's arranged in order. To find the median: (1) List the numbers in order from either largest to smallest or smallest to largest (you can pick); (2) If there is an odd number of values, the median is the value that has the same number of values above and below it on the list. (3) If there is an even number of values, the median is the mean of the two values right in the middle of the list.

10. Find the median test score on Exam 1.

Math Note

In Excel, the mean is calculated using the AVERAGE command. The median is calculated with the MEDIAN command, and the mode with the MODE command. Go figure. The syntax is just like the SUM command: enter =AVERAGE(B2:B14) to find the mean for Exam 1 above.

The **mode** of a list of numbers is the value that appears most often. If all values appear only once, there is no mode.

11. Find the mode of the Exam 1 scores.

At the beginning of the course, you should have been provided some information on how exactly you're going to be graded. Most instructors have their own standards, so you should make an effort to understand how your grade will be calculated. Most likely it was on a syllabus provided by your instructor.

12. How well do you feel you understand the grading standards for your course? If you don't know them, consult the syllabus or ask your instructor.

Let's look at a fairly basic points system for grading. We'll say that your course grade comes from 4 exams worth 100 points each, a homework score worth 200 points, and a final exam worth 200 points.

13. How many total points can be earned in the course?

14. What percentage of the total score is accounted for by the first exam?

15. If you don't show up for the first exam and take a 0%, then average 82% for everything else the rest of the course, what would your final percentage be?

	A	B	C	D	E	F	G	H	I
1		Exam 1	Exam 2	Exam 3	Exam 4	Homework	Final Exam	Total Points	Overall %
2	Points possible	100	100	100	100	200	200		
3	Your scores								

16. What would your final percentage be if you score 100% on Exam 1, then average 82% for the rest of the course?

	A	B	C	D	E	F	G	H	I
1		Exam 1	Exam 2	Exam 3	Exam 4	Homework	Final Exam	Total Points	Overall %
2	Points possible	100	100	100	100	200	200		
3	Your scores								

17. What does the difference between your answers to Questions 15 and 16 tell you?

1-10 Portfolio

Name ______________________________

Check each box when you've completed the task. Remember that your instructor will want you to turn in the portfolio pages you create.

Technology

1. ☐ Create a spreadsheet that calculates final grades for the grading system described on page 82. You can use the samples there as a template. Your sheet should use formulas to add up the number of points possible, your total score when you enter individual scores for exams and homework, and your final percentage. Experiment with your spreadsheet to find the effects of one unusually bad test score. A template to help you get started can be found in the online resources for this lesson.
2. ☐ In the online resources for Lesson 1-10, there's a video that describes how to make scroll bars. Enter the scores below for the four exams and homework, then put in a scroll bar with minimum value 0 and maximum value 200 to study the effects of the final exam score on your grade.

   Exam 1: 73 Exam 2: 78 Exam 3: 84 Exam 4: 82 Homework: 164

Skills

1. ☐ Include any written work from the online skills assignment along with any notes or questions about this lesson's content.

Applications

1. ☐ Complete the applications problems.

Reflections

Type a short answer to each question.

1. ☐ Write down a list of things that you think are most likely to affect how well you do on tests in this course. Then describe what you can do to make that information work for you.
2. ☐ There are many websites with test-prep and test-taking hints. Find one that you find helpful, and describe some tips that you can use.
3. ☐ Take another look at your answer to Question 0 at the beginning of this lesson. Would you change your answer now that you've completed the lesson? How would you summarize the topic of this lesson now?
4. ☐ What questions do you have about this lesson?

Looking Ahead

1. ☐ Read the opening paragraph in Lesson 2-1 carefully, then answer Question 0 in preparation for that lesson.

1-10 Applications Name ________________________________

Here's another look at a group of exam scores:

	A	B	C
1	**Student**	**Exam 1 (%)**	**Exam 2 (%)**
2	Michael	80	89
3	Andy	77	93
4	Pam	68	84
5	Jim	81	88
6	Dwight	96	91
7	Stanley	54	75
8	Phyllis	75	54
9	Kevin	81	86
10	Creed	71	0
11	Darryl	89	83
12	Gabe	56	64
13	Toby	81	65
14	Holly	92	73

1. Find the mean for all Exam 2 scores.

2. Find the median for all Exam 2 scores.

3. Find the mode for all Exam 2 scores.

4. Find the mean, median, and mode for the scores on Exam 2 if you throw out Creed's zero (dude didn't even show up ... Come on, man!).

5. You've found six measures of average for the Exam 2 scores so far on this page. Of the six, which single one do you think is the best representation of how students performed on the test? Needless to say, you should explain why you chose that one.

Unit 2
Relationships and Reasoning

Outline

Lesson 2-1 What Are the Chances?

Learning Objectives

- ☐ 1. Compute and interpret basic probabilities.
- ☐ 2. Express probability as a percent chance.
- ☐ 3. Understand the impact of events not being equally likely.

The 50-50-90 rule: Anytime you have a 50-50 chance of getting something right, there's a 90% probability you'll get it wrong.
– Andy Rooney

I'm planning on playing golf tomorrow, and my trusty Weather Puppy app tells me that there's a 10% chance of rain. So what exactly does that mean? Will it rain for 10% of the day tomorrow? Actually, a forecast like that is really just an educated guess: the forecaster is saying that there's about a one in ten chance that it will rain at some point tomorrow, which means I'll probably stay dry on the golf course. For our purposes, "probably" is the key word in the last sentence. That word indicates a certain likelihood that something will occur. In this lesson, we'll study *probability*, which is a way to assign a number or percentage to the likelihood of something occurring, like rain in Fairfield, OH tomorrow.

0. After reading the opening paragraph, what do you think the main topic of this section will be?

2-1 Group

If it doesn't make you uncomfortable, exchange the following information with the classmates in your Unit 2 group. This will be your small group for the second unit. It would be a good idea to schedule a time for the group to meet to go over homework, ask/answer questions, or prepare for exams. You can use this table to help schedule a mutually agreeable time.

Name	Phone Number	Email	Available times

The **probability** of an event occurring is a description of how likely it is that the event will actually happen. Probability can be described using a number ranging from 0 to 1, a percentage between 0% and 100%, or using words like impossible, unlikely, even chance, likely, or certain.

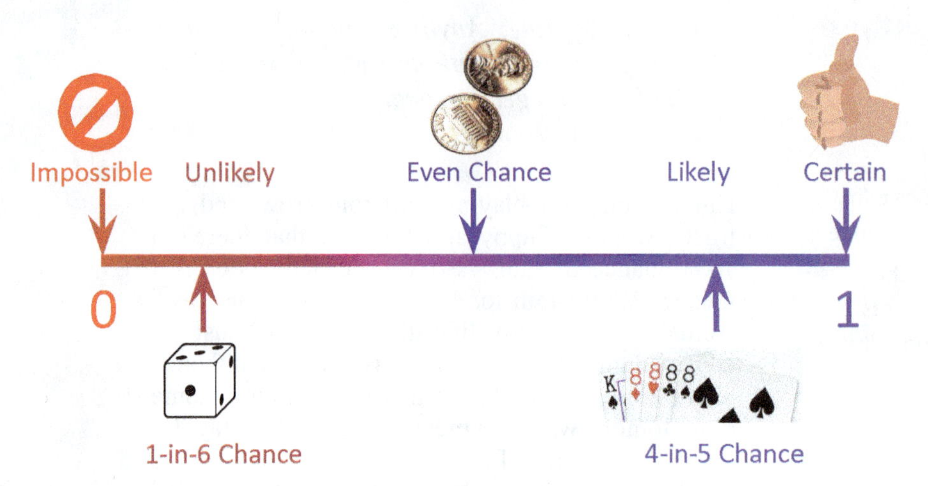

1. The probability of rolling a 4 with one standard six-sided die is $\frac{1}{6}$ because there is one side with 4 dots, and six sides total that could be showing when the die stops. Write this probability in equivalent forms as a decimal (rounded to three places) and as a percentage (rounded to one decimal place).

2. Write a sentence describing the percent chance of rolling a 4 with one standard die.

3. If you rolled a die 300 times, how many times would you expect it to land on 4?

4. How sure are you that you'd get the number of 4s you'd expect in 300 rolls? Explain.

5. When a standard coin is flipped, how many different outcomes are possible in terms of which side lands face up?

6. Based on your answer to Question 5, what is the percent chance that a single coin lands on heads when flipped?

7. Write the probability from your answer to Question 6 as a fraction between zero and one, and describe the likelihood verbally.

8. If you flipped a penny 80 times, about how many times would you expect it to land heads up?

9. Would you definitely get the number of heads you found in Question 8? Discuss.

10. Our probability diagram contains a picture of a good hand in poker: four 8s. If you were to randomly pick a card from that hand, what's the probability that it would be an 8? Write as a fraction between zero and one and as a percent.

11. Write a sentence describing the percent chance of drawing the king from that hand of cards.

Each of the probabilities that we've calculated so far can be found without doing any actual experiments: we know for sure that when we roll a die, there are six possible results, and only one of them is a 4. This type of probability is known as **theoretical probability.** Our calculations are based on observing how many outcomes are possible, how many fit a certain criteria, and assuming that each of the outcomes is as likely as any other to occur. If the die were weighted strangely to make 4 more likely to come up than other numbers, our probability calculation wouldn't be accurate anymore.

In that case, we might turn our attention to **empirical probability.** We could roll the die a bunch of times – let's say 100 – and record how many times 4 comes up. If we then divide the number of times we got 4 by the total number of rolls, that would give us an approximate percent chance of rolling 4.

Empirical Probability Lab

Supplies needed: A standard playing card, a dime, and scotch tape.
The point: If you flip a standard playing card and let it fall to the floor, the theoretical probability of it landing face up is 1/2. We can change the probability by making one side heavier; then it would make sense to compute an empirical probability.
The procedure: Tape the dime to the back side of the playing card. Then flip the card from eye level or higher, letting it fall to the floor without hitting anything. Note whether or not the front of the card lands face up. Repeat as many times as you think you need to in order to get a reasonable probability; any less than 20 would be a pretty bad idea. Record your results in the table using tally marks (|):

Number of flips	Times landing face up

12. Based on the results from your table, what is the percent chance that the card will land face up?

13. Write the percent chance as a probability between zero and one, then write a sentence describing how likely you think it is that the card will land face up.

14. Did you find the results of this experiment surprising based on how the card was altered? Explain.

15. Why was empirical probability a better idea than theoretical probability for this experiment?

Math Note

Experts have estimated that the probability of being struck by lightning is three times greater than the probability of winning $1 million or more in a lottery.

2-1 Class

1. What's wrong with the logic displayed by Rachel in the following conversation? Explain in depth.

Rachel: I have a 50-50 chance of passing this exam.
Ross: Why do you say that? That doesn't sound very optimistic.
Rachel: Well, it's just like flipping a coin – there are two possibilities. I'll either pass or I won't. So I have a 50% chance of passing.

2. Make a list of some things you can do to change the probability of passing an exam.

3. There are 45 applicants for 2 really great jobs, and you're one of them. If every applicant has an equal chance, what's the probability that you'll get the job?

4. Explain why theoretical probability is probably not a very realistic way of deciding your chances of landing the job in Question 3.

5. The promotional materials for a certain program at a community college boast that out of their 480 most recent graduates, 450 are currently working in the field. If you graduate in this program, describe what you think your chances are of getting a job. Include as many aspects as you can think of.

In many cases, you can use the results of existing surveys to compute empirical probabilities, eliminating the need to do an entire experiment on your own. Here's an interesting (or frightening, maybe) example: Jockey International surveyed 1,001 men to find the age of their oldest pair of underwear. The results are summarized in the table.

Age of oldest pair	Number of men
Less than 1 year	170
1-4 years	591
5-9 years	150
10-19 years	70
20 or more years	20

6. What's the probability that a randomly chosen man has no underwear older than 1 year?

7. What's the percent chance that a randomly chosen man has had a pair of underwear for over a decade?

8. Write a sentence or two explaining how likely you think it is that the oldest pair of underwear owned by the guy sitting closest to you in class is between 1 and 9 years old.

9. As we pointed out, 1,001 men were surveyed. The survey was done online. How accurate do you think the probabilities computed using this survey are when used to describe the habits of all American men? Discuss.

2-1 Portfolio

Name ______________________________

Check each box when you've completed the task. Remember that your instructor will want you to turn in the portfolio pages you create.

Technology

1. ☐ Find the results of an online survey that you find interesting, and use them to compute some empirical probabilities. Summarize your findings in a paragraph.

Skills

1. ☐ Include any written work from the online skills assignment along with any notes or questions about this lesson's content.

Applications

1. ☐ Complete the applications problems.

Reflections

Type a short answer to each question.

1. ☐ If someone asked you "What is probability, and why is it called that?", how would you answer?
2. ☐ Describe the relationship between probability and percent chance.
3. ☐ Take another look at your answer to Question 0 at the beginning of this lesson. Would you change your answer now that you've completed the lesson? How would you summarize the topic of this lesson now?
4. ☐ What questions do you have about this lesson?

Looking Ahead

1. ☐ Read the opening paragraph in Lesson 2-2 carefully, then answer Question 0 in preparation for that lesson.

2-1 Applications Name ____________________

1. According to the all-knowing Wikipedia, about 10% of all humans are left-handed. What would you say is the probability that the next person you run into is left-handed? What's the probability that he or she is right-handed?

2. In an average year, about 650 different pitchers pitch to at least one batter in a major-league baseball game. Based on your answer to Question 1, how many would you expect to be left-handed?

3. In reality, on average 175 of those pitchers are left-handed. What do you think that statistic says about being left-handed as a baseball pitcher? Explain.

4. In one math class, there are 30 students, 12 of whom are "non-traditional" in terms of age. If you randomly pick one person from that class, what's the probability that he or she will be of non-traditional age?

5. An online simulator helps students to understand their chances of passing a course based on study habits, attitude, attendance, work hours, sleep hours, and amount of homework completed. One students runs the simulation ten times, with results shown in the table. Based on these results, discuss how likely you think it is that the student will pass. More detail is better.

Simulation	Result
1	Fail
2	Pass
3	Pass
4	Pass
5	Pass
6	Fail
7	Pass
8	Pass
9	Fail
10	Pass

6. What do you think is the probability that you'll pass this course? Is this a set value, or can it change? If so, how?

Lesson 2-2 Of Planes, Boats, Doll Houses, and Dr. Evil

Learning Objectives

- ☐ 1. Understand the meaning of scale in models and maps.
- ☐ 2. Convert units using dimensional analysis.
- ☐ 3. View percentages in terms of scale.

I shall call him ... Mini Me.
– Dr. Evil

Have you ever thought about the number of different ways we have to measure objects? Inches, feet, yards, miles, centimeters, meters, kilometers ... and these cover only length; there are many more ways to measure length, as well as different units of measure for temperature, time, area, volume, mass, and other quantities. Why so many different measurements? The short answer is efficiency. Things come in so many different sizes that being stuck with just one unit of measurement would be very inconvenient. For example, it's about 5,892,480 inches from Chicago to Milwaukee. Clearly, measuring the distance in miles (93) is far more efficient. On the other hand, our friend Mini Me is 0.000505 miles tall, so measuring his height in inches (32) makes far more sense. In this section, we'll consider scale models (like Mini Me) as a way to learn an important skill: converting measurements from one unit to another.

0. After reading the opening paragraph, what do you think the main topic of this lesson will be?

THE MAIN TOPIC WILL BE TO CONVERT ONE UNIT TO ANOTHER

2-2 Group

1. Most people who pay any attention to the news are aware that the U.S. Military uses drones: unmanned aircraft that patrol the skies over unstable regions equipped with video and other surveillance equipment, and even weapons in some cases. These drones come in many different sizes, and the **scale** of a drone indicates its size compared to a similar full-sized plane. For example, a half-scale drone is half the size of a full-sized plane. List some other examples of where you've heard the term "scale" used in this way.

MODEL TRAINS,
MODEL CARS
MODEL BUILDING
DOLLS

The best gift I got for my birthday last year was my very own Mini Me; not a living one, which would have been the best gift that *anyone* ever got for their birthday, but a bobblehead doll made to look like me. (In case you're wondering, I don't wear shorts that short to play golf.) The scale for many objects is given as a ratio of two numbers. Ratios can be written either as fractions, or using a colon. If a scale model is exactly half the size of the original, we could say the ratio of the smaller size to the larger is 1:2, or we could write that ratio as 1/2. My bobblehead is a 1:13 scale model, which means that the handsome little devil is 1/13th of my actual height.

2. My Mini Me is 5.82 inches tall. How tall am I? Round to two decimal places.

$\frac{5.82\text{ IN}}{1} \cdot \frac{13\text{ REAL}}{1\text{ MINI}} = 75.66''$

3. If the actual Mini Me (32 inches tall) had a bobblehead made at the same scale, how tall would it be? Round to two decimal places again.

$\frac{32\text{ IN}}{1} \cdot \frac{1\text{ MINI}}{13\text{ REAL}} \quad \frac{32}{13} = 2.46\text{ IN}$

4. Write a verbal description of the process for going from larger to smaller and smaller to larger when given a scale, as in Questions 2 and 3.

LARGER TO SMALLER = DIVIDE
SMALLER TO LARGER = MULTIPLY

Here's a look at a map of the east side of Dubuque, Iowa. It shows a driving path from a casino on the river to a golf course on the other side. According to the maps app on my phone, the driving distance is 3.0 miles. We're most interested in the graphic at the bottom left, which indicates that 1 inch on the map (measure if you don't believe me) corresponds to 2,000 feet in real life.

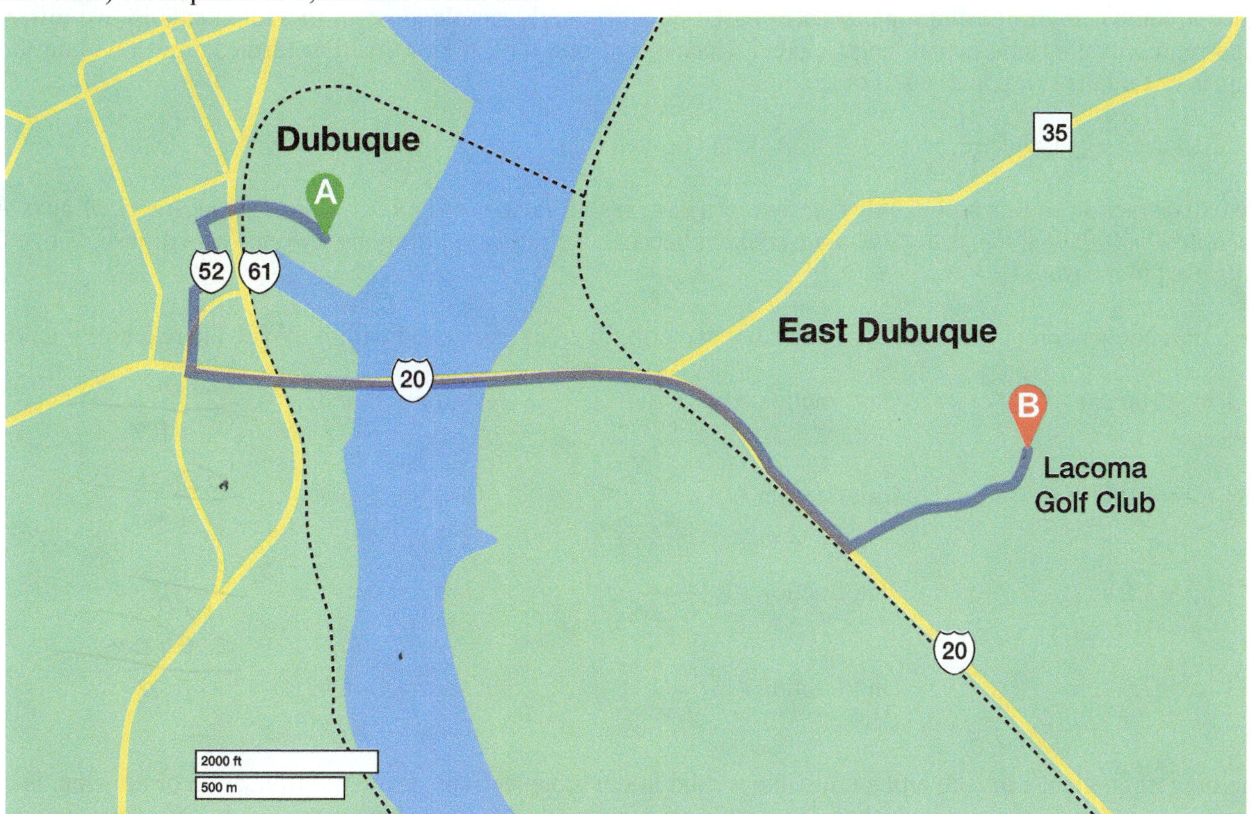

5. Points A and B are 4.06 inches apart on the map. How far is the casino from the golf course in feet and miles? (The straight-line distance, not the driving distance, which we already know.) Round miles to two decimal places.

$\frac{4.06 \text{ IN}}{1} \cdot \frac{2000 \text{ FT}}{1 \text{ IN}} = 8120 \text{ FT}$ $\quad \frac{8120 \text{ FT}}{1} \cdot \frac{1 \text{ MILE}}{5280 \text{ FT}} = \frac{8120}{5280 \text{ FT}}$ 1.54 MI

6. If we looked at a satellite photo at the same scale, could we see any boats on the river? If a 50-foot boat were on the river, how big would it appear on the photo? Round to three decimal places. Do you think you'd be able to see it?

$\frac{50 \text{ FT}}{1} \cdot \frac{1 \text{ IN}}{2000 \text{ FT}} \quad \frac{50}{2000} = .025 \text{ IN}$ NO!

7. Write a verbal description of the process of going from larger to smaller or smaller to larger in Questions 5 and 6.

LARGER TO SMALLER = DIVIDE
SMALLER TO LARGER = MULTIPLY

2-2 Class

Our next goal is to simplify the method for converting a measurement from one unit to another. This useful process, known as **dimensional analysis,** is based on a brilliantly simple idea: if measurements in two different units represent the same actual length, like 12 inches and one foot, then a fraction formed from dividing those units is just a fancy way to say 1. For example,

$$\frac{12\text{ in.}}{1\text{ ft}} = 1 \text{ and } \frac{1\text{ ft}}{12\text{ in.}} = 1$$

We call a fraction of this type a **unit fraction** or a **conversion factor,** and *multiplying any measurement by a unit fraction won't change the size of the measurement* because we're just multiplying by one; it will just change the units used to measure.

Each unit conversion in Questions 1–4 can have two unit fractions associated with it. Write the two unit fractions.

1. 1 m = 100 cm Unit fraction #1: Unit fraction #2:

2. 1 mi = 5,280 ft Unit fraction #1: Unit fraction #2:

3. 1 yd = 3 ft Unit fraction #1: Unit fraction #2:

4. 1 in. = 2.54 cm Unit fraction #1: Unit fraction #2:

The biggest challenge in converting units like we did in the group activity, using multiplication or division, is that it's easy to get confused over whether you should multiply or divide. Our clever procedure for converting units will be to multiply by one or more unit fractions in a way that the units we don't want will divide out and leave behind the units we do want, taking the guesswork out.

For example, to convert 22 feet to inches, the goal is to eliminate feet and leave behind inches. If we multiply by the unit fraction 12 in./1 ft, notice that the units "feet" appears in both numerator and denominator. You can think of this as having the length 1 ft in both the numerator and denominator. Of course, that 1 ft / 1 ft reduces to 1, leaving the units "inches". But since we ultimately just multiplied the original measurement by one (in the form of a unit fraction), the size of the measurement remains the same, and we've converted it to inches.

$$22\,\cancel{\text{ft}}\cdot\frac{12\text{ in.}}{1\,\cancel{\text{ft}}} = 264\text{ in.}$$

Conversion factors are particularly useful for conversions that have more than one step.

5. Let's say we want to convert the 3 miles of driving distance from the map in the group activity to yards.

a. First use your conversion factor from Question 2 to convert 3 miles to feet.

b. Now use your conversion factor from Question 3 to convert the answer from Question 5a to yards.

Math Note

You can think of the conversion of 22 feet to inches that we did as the units we didn't want (feet) dividing out to leave behind the units we do want (inches).

c. Now convert 3 miles to yards directly in one multi-step calculation. (This is the most efficient approach.)

$\frac{3 \text{ miles}}{1} \cdot \frac{1760 \text{ yards}}{1 \text{ mi}} = 5280 \text{ yd}$

6. We studied a variety of conversion factors in Lesson 1-9; use the table on page 74 to find how many cups are in a 2-liter bottle of Dr. Pepper. Round to two decimal places.

$\frac{2 \text{ lit}}{1} \cdot \frac{1.06 \text{ qts}}{1 \text{ lt}} \cdot \frac{2 \text{ pints}}{1 \text{ qt}} \cdot \frac{2 \text{ cups}}{1 \text{ pint}} = 8.48 \text{ cups}$ SENTANCE

Sometimes scale is described in terms of percentage. For example, for the filming of the Academy Award-winning film *Titanic*, a model of the stern of the ship was created at 12.5% of the actual size.

7. A section of rail on the stern of the real Titanic was 24 feet long. How long was the corresponding section on the model used for the movie?

$12.5\% = \frac{.125}{1000} \cdot \frac{24 \text{ ft}}{1} = \frac{3000}{1000} = 3$

8. If an I-beam on the model spanned 30 feet of deck space, how big was the corresponding I-beam on the actual ship?

$\frac{30 \text{ ft}}{1} \cdot \frac{1}{.125} = 240 \text{ ft}$

$\frac{30}{.125} = 240 \text{ ft}$

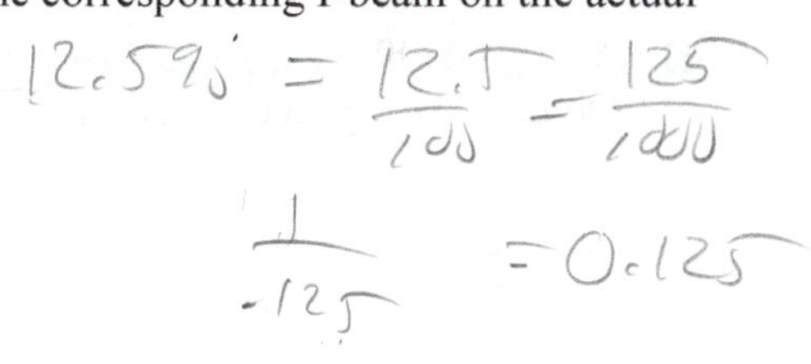

$12.5\% = \frac{12.5}{100} = \frac{125}{1000} = 0.125$, $\frac{1}{.125}$

9. Now let's turn this into a problem we already know how to solve: given that the model's size was 12.5% of the real ship's size, write a reduced fraction that describes the scale in fractional form, like we used earlier. (Hint: Recall that "percent" means "per hundred.")

$\frac{12.5\%}{100} = \frac{.12.5}{100}$

Quick quiz: Is it illegal to make color copies of United States currency? Most people think it is, but that's not entirely true. According to the Counterfeit Detection Act of 1992, you can make color copies as long as they're less than 75% or more than 150% of the original size.

10. Write a reduced fraction (like you did in Question 9) to describe each of those scales in fractional form.

$\frac{75}{100} = \frac{3}{4}$

$\frac{150}{100} = \frac{15}{10} = \frac{3}{2}$

11. An actual bill measures 2.61 in. by 6.14 in. Are we in trouble with the Feds for the reproduction above? Describe how you decided.

$\frac{6.14\text{ in}}{1} \cdot \frac{1}{4.3\text{ in}} = \frac{6.14}{4.3} = 143\%$ YES

2-2 Portfolio

Name ________________________________

Check each box when you've completed the task. Remember that your instructor will want you to turn in the portfolio pages you create.

Technology

1. ☐ Complete the following spreadsheet to calculate the lengths of models of different modes of transportation. Each model will be 1:16 scale. Use a formula that calculates the length of each model in feet, then another formula to convert those lengths to inches. A template to help you get started can be found in the online resources for this lesson.

	A	B	C	D
1		True Length (ft)	Model Length (ft)	Model Length (in)
2	Boat	17		
3	Car	20		
4	Truck	24		
5	Tractor	34		
6	Train	330		

Skills

1. ☐ Include any written work from the online skills assignment along with any notes or questions about this lesson's content.

Applications

1. ☐ Complete the applications problems.

Reflections

Type a short answer to each question.

1. ☐ Describe the process we developed for converting units; focus on the fact that we're multiplying by one.
2. ☐ Why do maps (like the one on page 97) have a scale in one corner?
3. ☐ Take another look at your answer to Question 0 at the beginning of this section. Would you revise your answer now that you've completed the lesson? How would you summarize the topic of this lesson now?
4. ☐ What questions do you have about this lesson?

Looking Ahead

1. ☐ Read the opening paragraph in Lesson 2-3 carefully, then answer Question 0 in preparation for that lesson.

2-2 Applications

Name WILLIAM BEZZER

1. Forensic scientists often examine scaled-up crime scene photos to search for evidence that might be hard to see at regular magnification. Studying bite marks on photo of a victim's leg at 500% magnification, a technician measures the distance between two punctures from the bite to be 165 mm. What is the actual distance between the punctures?

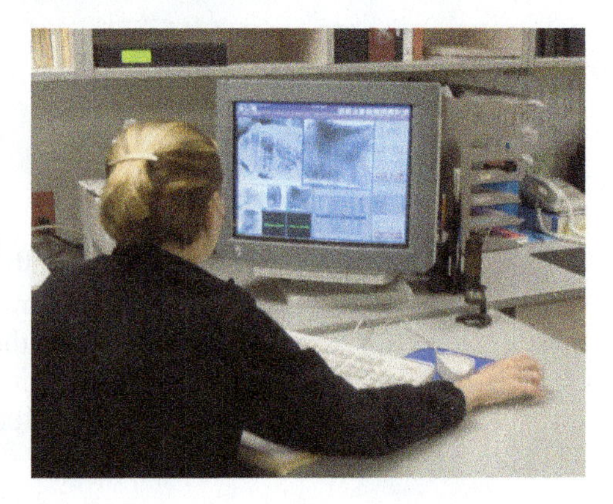

THE DISTANCE BETWEEN THE PUNCTURES IS 33 MM.

$\frac{165\text{ mm}}{5} = 33\text{ mm}$ $\quad \frac{1}{5} \cdot \frac{165}{1} = \frac{165}{5} = 33\text{ mm}$

2. In the 1984 movie *Ghostbusters*, a 112.5-foot-tall Stay Puft marshmallow man goes on a murderous rampage in New York City. After the film's successful box office run, a 20-inch replica of the big guy was sold in toy stores. What scale was used to make the replica? Give your answer both as a reduced fraction or ratio and as a percentage.

$\frac{112.5\text{ ft}}{1} \cdot \frac{12\text{ in}}{1\text{ ft}} \cdot \frac{1}{20\text{ in}} = \frac{1350\text{ in}}{20\text{ in}} = 67.5 = 6750\%$

THE SCALE OF THE REPLICA IS $\frac{1}{67.5}$

Doll houses provide an excellent example of scale modeling. The 1/2 inch scale is very common for doll houses; this means that 1/2 inch on the doll house corresponds to 1 foot for life-sized objects. Make sure you clearly show all calculations and units as you answer Questions 3–5.

3. What is an equivalent ratio form for the 1/2 inch scale as described above?

$\frac{.5\text{ in}}{1\text{ ft}} = \frac{.5\text{ in}}{12\text{ in}} = \frac{1}{24}$

THE EQUIVALENT RATIO FOR THE 1/2 INCH SCALE IS $\frac{1}{24}$ OR 24 TO 1, 1/24th

4. The width of one doll house is 34 in. Assuming that it uses the 1/2 inch scale, what's the width of the full-sized house that it's modeled after?

$\frac{34\text{ in}}{1} \cdot \frac{24}{1\text{ in}} = 816\text{ in}$

THE FULL SIZE HOUSE IS 816 INCHES OR 68 FT WIDE

5. If you're building a model stove for the doll house in Question 4, how tall should it be if the standard height for a full-sized stove is 36 inches?

$\frac{36\text{ in}}{1} \cdot \frac{1}{24} = \frac{36}{24} = 1.5$

THE STOVE FOR THE DOLL HOUSE SHOULD BE 1 1/2 INCHES TALL

2-2 Applications Name WILLIAM BECKER

For Questions 6–9, use dimensional analysis with unit fractions to convert each measurement to the given units. You'll need unit equivalences from this section and from page 74 to build your unit fractions. Round to two decimal places if necessary.

6. How many miles is a 10 kilometer race?

$\frac{10K}{1} \cdot \frac{1M}{1.61K} = \frac{10K}{1.61} = 6.21MI$

THERE IS 6.21 MILES IN A 10 KILOMETER RACE.

7. How many feet is a 10 kilometer race?

$\frac{10K}{1} \cdot \frac{1M}{1.61K} \cdot \frac{5280FT}{1M} = \frac{52800FT}{1.61}$ 32795 FT

THERE IS 32,795 FEET IN A 10 KILOMETER RACE

8. The average offensive lineman in the National Football League weighs 141.3 kilograms. How many pounds is that?

$\frac{141.3KG}{1} \cdot \frac{2.2LB}{1KG}$ 310.86LB

THE AVERAGE OFFENSIVE LINEMAN IN THE NFL WEIGHS 310.86 POUNDS

9. (This one requires more than one unit fraction.) If one liter of freshwater weighs 2.2 pounds, how much does the water in this 22,000 gallon swimming pool weigh?

$\frac{1LT}{2.2LB} \cdot \frac{1GAL}{3.79LT} \cdot \frac{1GAL}{8.338LB}$

22,000 G
× 8.338LB
= 183,436LB

$\frac{1GAL}{8.338LB} \cdot \frac{1}{22,000GAL} =$ 183,436LBS

THE WATER IN THE SWIMMING POOL WEIGHS 183,436LBS

Lesson 2-3 88 Miles Per Hour!

Learning Objectives

- ☐ 1. Interpret and use rates of change.
- ☐ 2. Convert units involving rates.

The future is something which everyone reaches at the rate of 60 minutes an hour, whatever he does, whoever he is.
– C.S. Lewis

Anyone familiar with the classic movie *Back to the Future* instantly recognizes the significance of 88 miles per hour: it's the speed the DeLorean time machine needed to reach in order to travel through time. Speeds (like 88 miles per hour) are familiar examples of the topic of this lesson, **rates of change.** Our world is a dynamic, ever-changing place, so studying the rate at which things change is an excellent way to study the world around us. Fortunately, studying rates will tie together some of the skills we've already practiced in this course, meaning we're in a good position to make the most of this useful topic.

0. After reading the opening paragraph, what do you think the main topic of this section will be?

RATES OF CHANGE

2-3 Class

A rate is a ratio that compares two quantities. By rate of change, we mean a rate that compares the change in one quantity to the change in another. A speed like 88 miles per hour qualifies, because we can write it this way:

$$\frac{88 \text{ mi}}{1 \text{ hr}}$$

This quite literally compares a change in distance (88 miles) to a change in time (1 hour). We call it a unit rate because its denominator is 1.

1. List some unit rates you're familiar with.

UNIT RATE

MILES PER HOUR
FEET PER SECOND
DOLLARS PER GALLON
PRICE PER POUND
CALORIES PER DAY

2. List some rates that aren't unit rates.

PAY PER HOUR(S)
MILES PER HOUR(S)
PER 1/4 MILE

$ / HRS $ / 1/4 HR   NOT

3. Compare rates of change and conversion factors. How are they similar? How are they different?

UNIT RATE MUST HAVE A 1 IN THE DENOMINATOR
THEY ARE DIFFERENT AS ONE OF THEM DOES NOT HAVE A 1

2-3 Group

The word **equivalent** is an important word in math, and one that we'll use often in this course. Make sure you clearly understand what that word means. Look it up and discuss it with your group if necessary.

1. If you make $100 a day for doing a certain job, how much money would you make in a five-day work week?

$\frac{\$100}{1 \text{ DAY}} \cdot \frac{5 \text{ DAYS}}{1} = \frac{\$500}{}$ YOU WOULD MAKE $500 IN A FIVE-DAY WORK WEEK

2. Are $\frac{\$100}{\text{day}}$ and $\frac{\$500}{5 \text{ days}}$ equivalent rates? What does that mean? YES! BECAUSE THEY HAVE THE SAME FRACTIONAL EQUIVALENT

Question 2 demonstrates a useful fact about rates of change: they can be scaled up or scaled down to fit a given situation. Scaling up a fraction is done by writing an equivalent fraction (that is, one that has the same value) using larger numbers. Scaling down is writing an equivalent fraction with smaller numbers. These processes are done by either multiplying BOTH the numerator and denominator by the same number, or dividing BOTH the numerator and denominator by the same number.

3. Write a fraction equivalent to $\frac{60 \text{ mi}}{4 \text{ hr}}$ that has 2 in the denominator.

$\frac{30 \text{ MI}}{2 \text{ HR}}$

4. Write a fraction equivalent to $\frac{60 \text{ mi}}{4 \text{ hr}}$ that has 1 in the denominator.

$\frac{15 \text{ MI}}{1 \text{ HR}}$

5. Write a fraction equivalent to $\frac{\$320}{4 \text{ days}}$ that has 10 in the denominator. (Hint: First scale down, then up.)

$\frac{\$320}{4 \text{ DAYS}} = \frac{\$80}{1 \text{ DAY}} \cdot \frac{\$800}{10 \text{ DAYS}}$

6. If one bicyclist is pedaling at a rate of 40 miles in 3 hours, and another at a rate of 50 miles in 4 hours, which is faster? (Scaling either up or down can be used!)

$\frac{40 \text{ MI}}{3 \text{ HR}} = \frac{13.33}{1 \text{ HR}} = 13.33$ MILES PER HOUR

$\frac{50 \text{ MI}}{4 \text{ HRS}} = \frac{12.5 \text{ MI}}{1 \text{ HR}} = 12.5$ MILES PER HOUR

THE BYCYCLIST WHO RODE 40 MILES IN 3 HOURS IS FASTER

Most experienced runners get to a point where they can comfortably jog long distances at a consistent pace. This "pace," of course, is another way to say "rate of change," because speed is the rate at which distance changes compared to time. One particular runner jogs one lap around a 400-meter track in 2 minutes. In Questions 7–15, you can assume that the runner can maintain this pace for a long time.

7. For every ___2___ minutes, the runner jogs ___400___ meters.

Time (Min)	Distance (m)
2	400
4	800
6	1200
8	1600
10	2000
12	2400

8. Complete the table, continuing the pattern for times and finding the associated distances for this runner.

9. Notice that the rate at which the runner's distance changes is constant. When the rate of change is constant, a quantity illustrates what type of growth (that we encountered in Unit 1)?

LINEAR

10. Write the runner's rate as a fraction using meters in the numerator.

$\frac{400\text{ m}}{2\text{ min}}$

Math Note

Using rates of change to calculate sizes of quantities is one of the most important applications of dimensional analysis.

11. Write this rate as a unit rate in meters per minute.

12. Convert your rate from Question 11 to meters per second using dimensional analysis. (Round to two decimal places.)

$\frac{200\text{ m}}{1\text{ min}} \cdot \frac{1\text{ min}}{60\text{ sec}} \quad \frac{200\text{ m}}{60} = 3{,}33$ METERS PER SECOND

13. What distance will the runner cover in 40 minutes?

$\frac{200\text{ m}}{1\text{ min}} \cdot \frac{40\text{ min}}{1} = 8{,}000$ METERS

14. What distance will he cover in 40 seconds?

$\frac{3{,}33\text{ m}}{1\text{ sec}} \cdot \frac{40\text{ sec}}{60\text{ sec}} = 133{,}2\text{ m}$

15. What would this runner's total time be for a 10k race? (Ten kilometers, that is. Recall that 1 km = 1,000 m. Your answer to Question 11 will help.)

$$\frac{10K}{1} = \frac{10{,}000\text{ m}}{1\text{ min}} \quad \frac{1\text{ min}}{200\text{ m}} \quad \frac{10000\text{ min}}{200\text{ m}} = 50\text{ min}$$

As Question 15 shows, when dealing with rates, sometimes it's necessary to convert the units in both the numerator and denominator of a fraction. (For example, it's common to convert speeds from feet per second to miles per hour; both feet and seconds need to be converted.) Fortunately, dimensional analysis is a perfect tool for this purpose, because the conversion factors we use equal one regardless of which unit is in the numerator.

16. The earth moves at about 98,000 feet per second as it revolves around the sun. How fast is that in miles per hour? (Recall that 1 mile is 5,280 feet.)

$$\frac{98{,}000\text{ ft}}{1\text{ sec}} \cdot \frac{1\text{ mi}}{5280\text{ ft}} \cdot \frac{60\text{ sec}}{1\text{ min}} \cdot \frac{60\text{ min}}{1\text{ hr}}$$

$$\frac{352800000}{5280} = 66{,}818.18182\text{ mph}$$

2-3 Portfolio

Name ______________________________

Check each box when you've completed the task. Remember that your instructor will want you to turn in the portfolio pages you create.

Technology

1. ☐ When you know the rate at which a quantity changes, you can build a spreadsheet to calculate the size of that quantity. Build a spreadsheet that calculates pay given hours worked and hourly rate, like the one shown here. You should be able to enter any rate you like in dollars per hour, and have a formula calculate the pay corresponding to any time you enter. Also enter a formula to calculate total pay for all employees. A template to help you get started can be found in the online resources for this lesson.

D11 fx

	A	B	C	D
1	Employee	Time (hrs)	Rate ($/hr)	Pay ($)
2	Napoleon	10	$8.75	$87.50
3	Deb	20		
4	Rico	32		
5	Me	40		
6			Total Pay	
7				

Skills

1. ☐ Include any written work from the online skills assignment along with any notes or questions about this lesson's content.

Applications

1. ☐ Complete the applications problems.

Reflections

Type a short answer to each question.

1. ☐ In this section, we learned that the speed of an object is a rate of change; specifically, the rate at which distance traveled changes as time changes. Think of at least three other rates of change, and describe specifically what they measure. Extra points if you come up with one that doesn't have a unit of time in it.
2. ☐ Why does dimensional analysis work for calculations involving rates? (Recall that we developed it as a method for converting units.)
3. ☐ Take another look at your answer to Question 0 at the beginning of this lesson. Would you change your answer now that you've completed the lesson? How would you summarize the topic of this lesson now?
4. ☐ What questions do you have about this lesson?

Looking Ahead

1. ☐ Read the opening paragraph in Lesson 2-4 carefully, then answer Question 0 in preparation for that lesson.

8

2-3 Applications

Name WILLIAM BERGER

When advertising special financing deals, car makers often use rates to describe what your monthly payment will look like. (This is because the amount you need to borrow varies widely based on the car you choose and the down payment you make.) In January 2013, Hyundai offered 2.9% financing on the Genesis coupe which would result in monthly payments of $22.09 per $1,000 borrowed for 48 months.

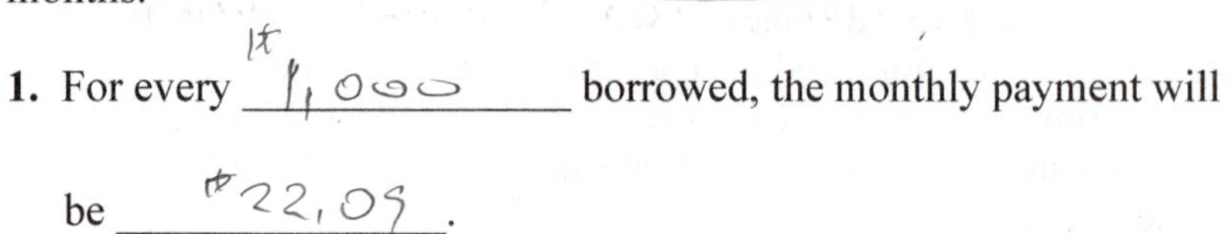

1. For every $1,000 borrowed, the monthly payment will be $22.09.

2. Complete the table of loan payments using your answer to Question 1.

Amount Borrowed	Loan Payment
$1,000	$22.09
$2,000	$44.18
$3,000	$66.27
$4,000	$88.36
$5,000	$110.45
$6,000	$132.54
$7,000	$154.63
$8,000	$176.72
$9,000	$198.81
$10,000	$220.90

3. Write the rate Hyundai advertised as a fraction.

$\frac{2.9\%}{100} = \frac{29}{1000}$ $\frac{\$22.09 \text{ PAYMENT}}{1000 \text{ BARROWED}}$

Use your answer to Question 3 and dimensional analysis for Questions 4 and 5.

4. What would your monthly payment be if you borrowed $18,000?

$\frac{18{,}000 \text{ BARROWED}}{1} \cdot \frac{22.09 \text{ PAY}}{1000 \text{ BAR}} = \frac{397.62}{1000} = 397.62$

MY MONTHLY PAYMENT WOULD BE $397.62

5. If your monthly payment is $331.35, how much did you borrow?

$\frac{331.35 \text{ PAY}}{1} \cdot \frac{1000 \text{ BAR}}{22.09 \text{ PAY}} = \frac{331350}{22.09} = \$15{,}000$

I BARROWED $15,000

6. If you agree on a price of $23,900 (including taxes and fees) for a Genesis and the dealership offers you $7,500 in trade for your old car, how much would your monthly payment be?

$23,900 − $7,500 = 16,400

$\frac{\$16{,}400 \text{ BAR}}{1} \cdot \frac{\$22.09 \text{ PAY}}{\$1000 \text{ BAR}} = \frac{\$362276}{1000}$ $362.28

MY MONTHLY PAYMENT WOULD BE $362.28

2-3 Applications Name ________________

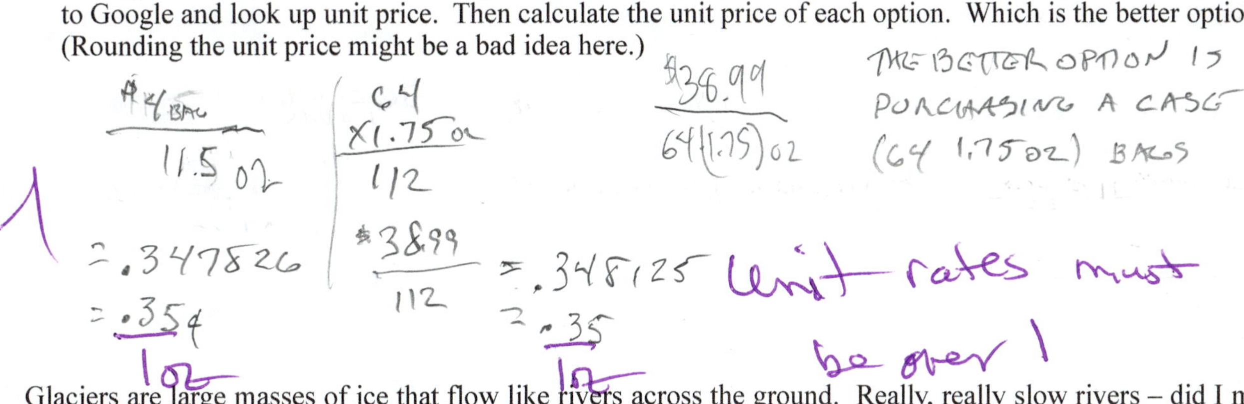

7. When shopping for Doritos for your large office party, you find that you can buy several 11.5 oz bags for $4 each or a case of 64 small 1.75 oz bags for a total of $38.99. Always trying to save money, your boss says, "Just get the one with the lower unit price." What should you do? First, if you need to, you should go straight to Google and look up unit price. Then calculate the unit price of each option. Which is the better option? (Rounding the unit price might be a bad idea here.)

Glaciers are large masses of ice that flow like rivers across the ground. Really, really slow rivers – did I mention that they're ice? Most move less than a foot per day. At one point, the San Rafael glacier in Chile was moving 203 millimeters per day.

8. How fast was it moving in inches per hour? (1 inch is 25.4 mm.)

9. Find the speed in miles per hour, then explain why that's a silly unit of speed in this case. (Note that your calculator will probably give you an answer in scientific notation. If you're not familiar with that, a little Internet research might be in order.)

10. How long would it take the glacier to move the length of a football field (100 yards)? (Round to the nearest tenth of an hour.)

Maribel the reason that make River's family came to the United States. Next to the stories of the main characters are the stories of the neighbors. One of the stories is about Quisqueya Solis. She is from Venezuela. She is a single mother who spends her time for two boys. Everything about her becomes more meaning when she has those good sons who go to university. Another single mother in the novel is Alma. At the end of the journey, her husband died. Now everything depends on her. From now she has to take care Maribel by herself without helping from her husband. Her husband Arturo used to decide everything. He decided to bring his family to the United States, got a job to take care his family, and he is a person who tries to make Alma who always thinks that Maribel's accident is her mistake feels better. When he

$$\frac{100\ \text{YDS}}{} \cdot \frac{3\ \text{FT}}{1} \cdot \frac{12\ \text{IN}}{1\ \text{FT}} \cdot \frac{1\ \text{HR.}}{.33}$$

Lesson 2-4 It's All Relative

> **Learning Objectives**
>
> ☐ 1. Compare difference to relative difference.
>
> ☐ 2. Apply relative error.
>
> ☐ 3. Find conversion factors for square and cubic units.

Put your hand on a hot stove for a minute, and it seems like an hour. Sit with a pretty girl for an hour and it seems like a minute. That's relativity.
– Albert Einstein

In a nutshell, Einstein's theory of relativity says that even the passage of time is not an absolute, but rather is relative to your frame of reference. In this lesson, we'll study appropriate ways to compare numbers, focusing on the importance of relativity. Consider these two scenarios: 1) I owe you \$20, and I give you \$10 instead. 2) You win \$5 million in a lottery, but you only get paid \$4,999,990. In which scenario was the shortage significant? You're out ten bucks in each, but clearly that's a much bigger deal when the amount owed is \$20. If that makes sense to you, then you have an instinctive understanding of the concept of relative difference.

0. After reading the opening paragraph, what do you think the main topic of this section will be?

2-4 Class

When asked to compare the sizes of two numbers, most people think of subtraction, but the example above shows that this isn't necessarily the best choice. The **difference** between two values is found by simply subtracting the new value from the old:

Difference = New value – Old value

In both scenarios above, the difference in what was owed and what was paid is \$10, which leads us to conclude the significance of that \$10 is the same in each case. But doesn't it seem clear that the \$10 shortfall is more significant when the total amount owed is \$20? That's where relative difference comes in. The **relative difference** between two values measures the difference as a fraction of the original value.

$$\textbf{Relative difference} = \frac{\textbf{Difference}}{\textbf{Original value}} = \frac{\textbf{New value} - \textbf{Original value}}{\textbf{Original value}}$$

In Scenario 1 above:

$$\text{Relative difference} = \frac{\$10 - \$20}{\$20} = -\frac{1}{2} \text{ or } -50\%. \text{ (The negative indicates that you got shorted. Bummer.)}$$

This shows that the difference between the amount owed and the amount received is half of the amount owed. That's a big deal!

> **Math Note**
>
> Relative differences are often written as percentages, but they don't have to be. Fraction form is fine, too.

1. Find the relative difference if you win \$5,000,000 and get shorted by \$10. Write as a fraction, not a decimal.

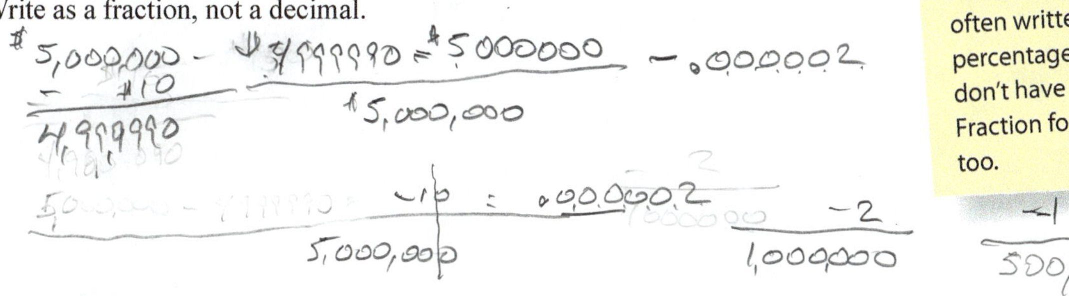

2. Compare the relative differences for the two scenarios in the opening paragraph from this lesson. Does this match your intuition about the scenarios?

YES IT MATCHES

3. Both Shawna and John found out that they're getting a $1,000 annual raise this year. John went out to celebrate, while Shawna yawned and said "Yeah, whatever." Use the topic of this section to describe what you think might be likely to account for this discrepancy in their reactions to the news.

I BELEIVE THE REASON FOR THE DISCREPANCY IS THAT JOHN ONLY MAKES $15,000 PER YEAR WHICH MAKE IT A BIG RAISE AND SHAWNA MAKES $85,000 PER YEAR AND THAT WOULD BE A SMALL RAISE

4. Your instructor tells you that you're going to get an additional 2 points on a recent test or quiz. Describe when this would be great news, and when it wouldn't help very much.

IT WOULD BE GREAT NEWS ON A QUIZ THAT IS WORTH 10 POINTS

IT WOULD NOT HELP MUCH ON A TEST THAT IS WORTH 100 POINTS

5. The U.S. Federal budget deficit for 2011 was about 1.298 trillion dollars. (This means that the government spent almost $1.3 trillion dollars more than it took in.) If you round that amount up to 1.3 trillion dollars, which seems perfectly harmless, you cost the taxpayers of this country 2 BILLION dollars. Thanks. What's the relative difference between the actual amount and the higher rounded amount? (Hint: If you're clever, you won't need to work with huge numbers...)

$\$1.3 - 1.298 = \frac{.002}{1.298} = .00154162$

$\frac{\$1.298 - 1.3}{1.3} = \frac{-.002}{1.3} = .001538462$

2-4 Group

One interesting application of relative difference is relative error. When some quantity is estimated, the difference between the actual amount and the estimated amount is the **actual error.** The **relative error** is the relative difference between those amounts, or the actual error divided by the actual amount.

If you're ever in the market for some new carpet, you might consider searching the Web for the Carpet Buyer's Handbook. The site has many useful resources on buying and installing carpet. It turns out that figuring out how much you're going to spend is more complicated then you might think. Carpet is sold by the square foot, and you can measure the square footage of a room. But you can't buy exactly a certain number of square feet because carpet comes on rolls of a certain width (usually 12 feet). If you need a 2 ft × 3 ft piece for a closet, you can't buy just 6 square feet: you'd have to buy 2 feet off of the roll, which would be 24 square feet.

In addition, there are certain guidelines that installers will adhere to in order to get the best result. These govern how the pieces off the roll can be situated, leading to more potential waste. The bottom line is that when a professional installer measures a room to estimate the amount of carpet needed, the result will usually be quite a bit more than the actual square footage that will be covered.

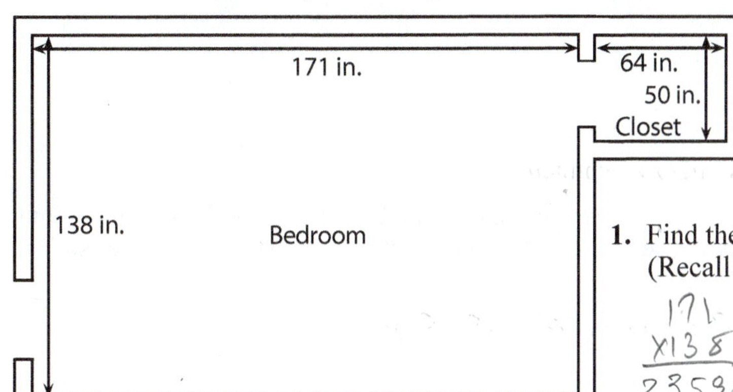

Using professional guidelines, an installer estimated that $24\frac{2}{3}$ square yards of carpet would be needed for the bedroom and closet shown in the diagram.

1. Find the area of the bedroom only in square inches. (Recall that the area of a rectangle is length times width.)

171
×138
23598

2. Find the area of the closet in square inches.

64
×50
3200

3. Each of the doorways shown in 32 inches wide, and the walls are 4 inches thick. Carpet is needed for the entire doorway between the bedroom and closet, and in half of the doorway into the bedroom. How many square inches does this add to the total carpeted area?

32 × 4 = 128   4 × .5 = 2   32 × 2 = 64   128 + 64 = 192

IT WOULD ADD 192 SQUARE INCHES TO THE TOTAL CARPETED AREA

4. We know that there are 36 inches in one yard. How many square inches are there in a square yard? The diagrams below should help.

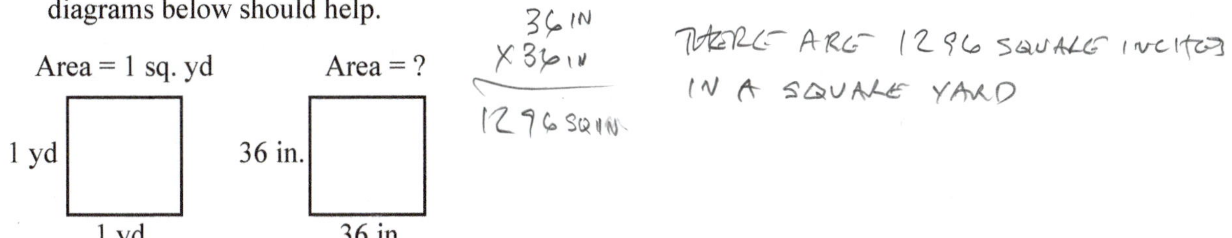

36 IN
× 36 IN
1296 SQ IN

THERE ARE 1296 SQUARE INCHES IN A SQUARE YARD

5. Use the conversion factor you developed in Question 4 to convert the total carpeted area to square yards. Round to one decimal place.

23598
3200
182
26,990 SQIN

36
X36
1296

26,990 / 1296 = 20.8 SQ YD

6. Write a sentence or two describing how the professional estimate compares to the actual carpeted area.

THE INSTALLER FIGURES HE WILL NEED 7 YARDS FROM A ROLL THAT IS 12 FEET WIDE

7. Find the relative difference between the estimate and the actual area of the carpeted portion of the room. Round to two decimal places. Write a sentence explaining what this tells us about how the estimate compares to the actual carpeted area. Writing the relative difference as a percentage is a good idea.

24.667 − 20.8 = 3.867 / 24.667 = .156 = 16%
= .16 =

THIS TELLS US THAT BECAUSE OF THE AMOUNT OF WAIST OF THE ESTIMATE IS APROX 16% MORE THAN THE ACTUAL MEASURMENTS

8. Not everyone realizes this, but when you buy carpet, you pay for the estimated amount, not the actual amount covering your floors. If the family that owns the house in the diagram picked carpet that costs $28 per square yard installed, how much would they have paid?

24.667 SQYD
X $28
$690.68

THEY WOULD HAVE PAID $690.68 FOR THE CARPET

9. Use the actual difference between the estimate and the carpeted area to find the dollar value of wasted carpet.

$690.68
X .156
$107.75

10. List some situations where relative difference is a much more useful number than actual difference.

NO!

2-4 Portfolio

Name ______________________________

Check each box when you've completed the task. Remember that your instructor will want you to turn in the portfolio pages you create.

Technology

1. □ Complete the following spreadsheet. Use formulas in all the cells in row 3, then copy them to fill out the remaining rows. Then write any observations you have based on the results. A template to help you get started can be found in the online resources for this lesson.

	A	B	C	D	E	F	G
1	Account #1	Difference	Relative Difference		Account #2	Difference	Relative Difference
2	$10,000.00				$10,000.00		
3	$10,700.00				$10,500.00		
4	$11,400.00				$11,025.00		
5	$12,100.00				$11,576.25		
6	$12,800.00				$12,155.06		
7	$13,500.00				$12,762.82		
8	$14,200.00				$13,400.96		
9	$14,900.00				$14,071.00		

Skills

1. □ Include any written work from the online skills assignment along with any notes or questions about this lesson's content.

Applications

1. □ Complete the applications problems.

Reflections

Type a short answer to each question.

1. □ Based on the spreadsheets you created in the Technology section, you can find a connection between the concept of actual vs. relative difference, and the concept of linear vs. exponential growth, which we studied in Unit 1. Write a paragraph describing that connection.
2. □ Account #2 in the Technology section is an example of compound interest. What is the interest rate? (Hint: Think about the interest in the first year.) Where does this rate appear in the spreadsheet?
3. □ How would you explain to someone you know that relative difference is often much more meaningful than actual difference?
4. □ Take another look at your answer to Question 0 at the beginning of this lesson. Would you change your answer now that you've completed the lesson? How would you summarize the topic of this lesson now?
5. □ What questions do you have about this lesson?

Looking Ahead

1. □ Read the opening paragraph in Lesson 2-5 carefully, then answer Question 0 in preparation for that lesson.

2-4 Applications

Name WILLIAM BEZZER

After moving into a new house, a couple wants to have a concrete patio poured to support a hot tub, because... well, because it's a hot tub. The plans call for a 14 ft by 16 ft slab of concrete 4 inches thick. Round any calculations to two decimal places if you need to.

1. How many cubic feet of concrete will be needed?

14 FT 16 X 12 IN X 12 IN — YOU WILL NEED 74.67 CUBIC FEET FOR THE PATIO

168 • 192 • 4 = 129,024 ÷ 1728 = 74.67 CU FT   12 • 12 • 12 = 1728 CU IN

2. Convert the volume of concrete to cubic yards. (Hint: A cube that is one yard on each side is also 3 feet on each side. You can use that to find a conversion factor from cubic feet to cubic yards.)

74.67 CU FT • 1 CU YD / 27 CU FT   74.67 / 27 = 2.77 CU YD

74.67 CU FT CONVERTS TO 2.77 CU YDS   3 X 3 X 3 = 27

3. The coolest thing about ordering ready-mix concrete is that a concrete mixer will show up at your house. Most companies require orders by the cubic yard. The patio above was estimated at 3 cubic yards. How much extra concrete was ordered?

3 CU YDS − 2.77 = .23 CUBIC YARDS

THERE WAS .23 CU YDS EXTRA CONCRETE ORDERED

4. Find the relative error in the estimate compared to the actual amount of concrete needed.

(3 CU YD − 2.77) / 2.77 = .23 / 2.77 = .083 = 8.3%

THERE WAS A RELATIVE ERROR OF 8.3%

5. In 2012, an average cost for concrete was $140 per cubic yard. How much would be spent on concrete for the patio at that price?

3 • $140 = $420

THERE WOULD BE $420 SPENT ON CONCRETE FOR THE PATIO

6. Find the dollar value of the wasted concrete.

$140 × .23 = $32.20

THERE WAS $32.20 VALUE WASTED ON EXTRA CONCRETE

Lesson 2-5 Ins and Outs

Learning Objectives

- ☐ 1. Distinguish between inputs (independent variables) and outputs (dependent variables).
- ☐ 2. Evaluate expressions and formulas.
- ☐ 3. Write and interpret expressions.

Experience is a hard teacher because she gives the test first, the lesson afterward.
– Vernon Law

The holy grail of energy research: a process that outputs far more energy than you put in. Develop a safe, clean process that does that, and you'll be rich beyond your wildest dreams. There's a worthwhile metaphor here: in many instances, the output (result you get) is determined by what you input. Don't practice and stay out all night before a big game, your output will probably stink. Put in a ton of work and effort, and you're likely to do your best. In this lesson, we'll study input and output from a mathematical sense, and hopefully give you a MUCH better idea of just what in the world a "variable" really is.

0. After reading the opening paragraph, what do you think the main topic of this section will be?

2-5 Class

First, let's talk about variables. Answer this question honestly and to the best of your ability without any outside help:

1. What is a variable?

If you're like most people, your answer was probably something like this: "A variable is a letter instead of a number." I am very sorry to inform you that, and I say this in the most caring possible way, your answer stinks. Think about what the word "variable" should mean in plain English: able to vary. And THAT is the key to understanding variables:

A **variable** is a quantity that is able to change, or vary.

Contrast this with a **constant,** which can't vary. The number 12 is a constant, because no matter what, it still has the same value. The number of hours you spend studying, on the other hand IS a variable, because it can vary depending on how much effort you choose to expend.

But what about the whole "letter" thing? Isn't x a variable? Technically, the answer is no. We use letters to REPRESENT variables, since that distinguishes them from numbers, which never change. But a variable is NOT a letter: it's a *quantity* that can vary which is usually represented by a letter. To save time, we often refer to the letter itself as a variable, but try to keep in mind it's actually the quantity represented by that letter that is actually a variable.

2. Which sentence makes more sense? Explain your reasoning.

a. The amount of time you spend studying depends on the grade you earn in a course.
b. The grade you earn in a course depends on the amount of time you spend studying.

3. List other factors that are likely to impact your grade in this course.

STUDING -

DOING ASSIGNMENTS

Because these factors can cause a change in your course grade, and not the other way around, the course grade (which can vary) is referred to as a **dependent variable** in this situation. Other factors, like the amount of time you spend studying, are called **independent variables.** Independent variables cause changes in dependent variables.

4. Consider the following relationships, where one quantity or event causes another to change. Identify the independent and dependent variable in each case, and don't forget to think about WHY it makes sense that these things are variables.

a. The age of a tree and the height of a tree

AGE OF TREE INDY

HEIGHT IS DEPEND

b. The number of practice sessions and the quality of a musical performance

NO PRACTICE = INDY

QUALITY DEPENT

c. Your score on a placement test and the math courses you've taken previously

SCORE IS DEPEND

MATH TAKE INDY

d. Your blood pressure and the amount of time you spend exercising each week

BLOOD DEPEND

EXERSISING INDY

e. The value of a share of Apple stock and what year it is

YEAR = DEPEND

VALUE = INDY

f. The number of songs a band sells on iTunes and the amount of money spent on marketing

SONGS = INDY

MONEY = DEPEN

g. The cost of a cab ride and the number of miles you're driven

miles = INDY

COST = DEP

h. The number of customers that want to buy a certain product and the price of that product

PRODUCT = INDY

CUSTOMER = DEP

2-5 Group

1. A few key terms used in this section are listed below. Discuss each term with your group and note how the definitions are similar, and how they're different.

Expression – A combination of variables and constants using mathematical operations and grouping symbols

Examples: $4x+2y$, $\frac{11}{t-2}$, $\sqrt{x^2-20}$, $(4n-6)(2+n)$

Equation – A statement that two quantities are equal built using expressions and an equal sign (=)

Examples: $3x+2=7$, $y=z^2-z-6$, $100e^{0.02t}=500$

Formula – An expression or equation with multiple variables that is used to calculate some quantity of interest

Examples: πr^2, $P=2l+2w$, $A=P+\text{Pr}t$

In many cases, relationships like the ones we thought about in the Class portion of this lesson can be described mathematically using a formula. For example, it might cost $2 for a cab ride plus $2.50 for each mile; in that case we could write the formula

$$C=2+2.50m$$

where C represents the variable cost of the cab ride, and m represents the variable number of miles. It's the number of miles that affects the cost, so m is the independent variable and C is the dependent variable. We might say that the number of miles is the **input,** while the cost is the **output.** Think of it like a machine: you input a number of miles, and the machine gives you back the cost.

2. If we wanted to know the cost of a 7-mile cab ride, we could replace the variable m, which represents miles; the result would be $C=2+2.50(7)$. Finding the value of the calculation to the right of the equal sign will tell us the cost. We call this **evaluating** the expression (or formula) for cost. Find the cost of the 7-mile cab ride.

C = 2 + 2.50m
C = 2 + 2.50(7)
C = 2 + 17.50
C = 19.50

Math Note

There's no reason you HAVE to use a letter to represent a variable quantity: You could use a smiley face, a zodiac sign, a picture of your mom, whatever. But we'll use letters because we don't know what your mom looks like.

3. Evaluate the cost formula for $m=4$, then attach units to your answer and write a sentence describing what it tells us. Include information about each variable.

C = 2 + 2.50m
C = 2 + 2.50(4)
C = 2 + 10
C = 12

The formula $A = P(1+r)^t$ gives the value A of an account where P dollars have been deposited for t years at a compound interest rate r (written in decimal form).

4. Evaluate this formula for $P = \$10,000$, $t = 40$, and $r = 0.06$.

~~A = 10000(1+r)^t~~
A = 10000 (1+.06)^40
A = 10000 • 1.06^40
A = 10,000 • 10.285717194
= 102857.18

5. Write a short paragraph describing the significance of the calculation in Question 4. What does it tell us? Include information about each of the variables.

IT TELLS US THAT IT IS A MUST TO USE THE ORDER OF OPPERATIONS, AND TO FIND THE INDEPENDANT AND DEPENDANT VARIABLES

Next, we're going to look at how the types of relationships we're studying can be applied to a small business. This spreadsheet describes the costs associated with running a snow cone business, as well as revenue and profit.

	A	B	C	D	E	F	G	H	I	J	K	L
1	Snow cones sold	Cups	Cost of cups	Syrup (oz)	Cost of syrup	Ice (oz)	Cost of ice	Combined supply costs	Fixed costs	Total costs	Revenue	Net profit
2	0	0	$0.00	0	$0.00	0	$0.00	$0.00	$245.00	$245.00	$0.00	-$245.00
3	100	100	$1.00	200	$4.00	500	$7.00	$12.00	$245.00	$257.00	$75.00	-$182.00
4	200	200	$2.00	400	$8.00	1000	$14.00	$24.00	$245.00	$269.00	$150.00	-$119.00
5	300	300	$3.00	600	$12.00	1500	$21.00	$36.00	$245.00	$281.00	$225.00	-$56.00
6	400	400	$4.00	800	$16.00	2000	$28.00	$48.00	$245.00	$293.00	$300.00	$7.00
7	500	500	$5.00	1000	$20.00	2500	$35.00	$60.00	$245.00	$305.00	$375.00	$70.00

6. If 300 snow cones are sold, how many ounces of syrup are needed?

IT WOULD NEED 600 OZ OF SYRUP

7. A formula was typed into cell D5 to obtain the result you used to answer Question 6. What formula would give the correct result? (The formula should use cell A5.)

= A5*2

8. Write a verbal description of the relationship between the number of snow cones sold and the number of ounces of syrup that will be needed.

YOU WOULD NEED 2 OZ OF SYRUP PER SNOW CONE

9. Use your answer to Question 8 to write an expression with variable *x* that represents the number of ounces of syrup used. The variable *x* should represent the number of snow cones sold.

2X

10. Using the letter *S* to represent the amount of syrup used, write a formula that describes the ounces of syrup needed in terms of the number of snow cones sold.

S = 2X

11. Does the number of snow cones sold depend on the amount of syrup used, or does the amount of syrup used depend on the number of snow cones sold?

THE AMOUNT OF SYRUP USED DEPEND ON THE NUMBER OF SNOW CONES SOLD

12. In your formula from Question 10, which is the independent variable, and which is the dependent variable?

THE INDEPENDENT VARIABLE IN THE SNOW CONES AND THE DEPENDENT IS THE SYRUP

13. If x represents the number of snow cones sold, what quantity described in the spreadsheet would be represented by $5x$? = ICE

5 · X = ICE

14. If 70 snow cones are sold, the value of $5x$ is 350. Write a sentence or two explaining exactly what that result tells us about the snow cone business.

IT TAKES 5 OZ OF ICE FOR EACH SNOW CONE SOLD

15. Write a unit rate that describes the combined costs of all supplies needed per snow cone sold.

$\frac{12}{100} = \frac{\$.12}{1}$ CONE

16. Notice that the fixed costs are \$245 for any number of snow cones sold. That's what makes them **fixed costs:** they aren't affected by the sales. The formula for finding total costs in terms of the number of snow cones sold (x) is $C = 0.12x + 245$. Explain how we got that formula.

WE GOT THAT FORMULA BY ADDING FIXED COST PLUS THE AMOUNT OF SNOW CONES SOLD

2-5 Portfolio

Name ______________________________

Check each box when you've completed the task. Remember that your instructor will want you to turn in the portfolio pages you create.

Technology

1. ☐ The spreadsheet on page 122 is in the online resources for Lesson 2-5. Make a copy of the spreadsheet, then use the copy down feature to extend the existing pattern down to show up to 2,100 snow cones sold. This should be in row 23 if all goes well.
2. ☐ Now create a scroll bar that changes the values in cell A23. (We first learned how to make scroll bars in the online resources for Lesson 1-10.) The settings should be Current = 2,000, minimum = 0, maximum = 2,000. Adjust the scroll bar to the point where the profit is as close to zero dollars as possible. Record the value on the scroll bar, and describe what that tells you about the business.

Skills

1. ☐ Include any written work from the online skills assignment along with any notes or questions about this lesson's content.

Applications

1. ☐ Complete the applications problems.

Reflections

Type a short answer to each question.

1. ☐ What is a variable? Why is it wrong to answer "a letter"?
2. ☐ When two variables are related, what does it mean to say that one is the independent variable and the other is the dependent variable?
3. ☐ Describe the difference between an equation and an expression.
4. ☐ How do the terms input and output apply to the relationships we discussed in this lesson?
5. ☐ Take another look at your answer to Question 0 at the beginning of this lesson. Would you change your answer now that you've completed the lesson? How would you summarize the topic of this lesson now?
6. ☐ What questions do you have about this lesson?

Looking Ahead

1. ☐ Read the opening paragraph in Lesson 2-6 carefully, then answer Question 0 in preparation for that lesson.

$5-(7-x)$
$5-7+x$
$-2+x$

2-5 Applications

Name WILLIAM BEZZER

Use the spreadsheet on page 122 to answer the following questions.

1. Find the total cost from selling 400 snow cones.
$293.00
THE TOTAL COST OF FROM SELLING 400 SNOW CONES IS $293.00

2. Find the revenue from selling 400 snow cones.
$300.00
THE REVENUE FROM SELLING 400 SNOW CONES IS $300

3. A formula that uses cells J7 and K7 was typed in cell L7. What is that formula? (Focus on the dollar amounts in cells J7 and K7.) = K7 − J7
THE FORMULA FOR L7 IS = K7 − J7

4. Write a verbal explanation of the relationship between the profit made from selling a certain number of snow cones and the revenue and total costs from selling that number of snow cones.
THE RELATIONSHIP BETWEEN THE PROFIT MADE IS, THE PROFIT IS THE AMOUNT FROM SUBTRACTING TOTAL COSTS FROM THE REVENUE

5. If we use *P* to represent profit, *C* to represent total costs, and *R* to represent revenue, write a formula that describes the profit in terms of *C* and *R*.
THE FORMULA FOR REPRESENTING PROFIT IS P = R − C
$P = R - C$

6. In Question 16 of the 2-5 group activity, you discussed how to represent total costs with the expression $0.12x + 245$. Using the same line of reasoning, find an expression that describes the revenue from selling *x* snow cones. (Recall that we used a unit rate to find the expression for total costs.)
$0.75 * X$
AN EXPRESSION THAT DESCRIBES THE REVENUE FROM SELLING X SNOW CONES IS 0.75 • X

7. Is $5 + (7 + 3)$ equal to $5 - 7 - 3$, or $5 - 7 + 3$? Perform each calculation to decide, then try to explain why it worked out that way.
$5-7-3 = -2-3 = -5$
$5-7+3 = -2+3 = 1$
$5-(7-3) = 5-4 = 1$
5−7+3 IS THE MORE ACCURATE CALCULATION BECAUSE IT GIVES THE SAME ANSWER AS THE NORMAL WAY OF 5 − (7−3)

8. Subtract the expression $0.12x + 245$ from the one you wrote in Question 6. Use what you discovered in Question 7 to perform the subtraction. Then write a description of what exactly this new expression represents.
$P = R - C$
$P = 0.75 \cdot X - (.12X + 245)$
$P = .75X - .12X - 245$
$P = .63X - 245$
WE ARE SUBTRACTING THE COST OF EACH SNOW CONE AND THE FIXED COST FROM THE REVENUE OF EACH SNOW CONE

Lesson 2-6 Oh Yeah? Prove It!

Learning Objectives

☐ 1. Use inductive reasoning to make a conjecture.

☐ 2. Disprove a conjecture by finding a counterexample.

☐ 3. Use deductive reasoning.

When you have eliminated the impossible, whatever remains, however improbable, must be the truth.
– Sherlock Holmes

Here's a hypothetical situation: you're in an English literature course, and the prof reads a passage from a Shakespearean sonnet, then says "Clearly, Shakespeare was referring to the forthcoming rise of reality television when he wrote that passage." Would you (a) mumble "Wow, no kidding" and write his claim down in your notebook, or (b) respectfully disagree and challenge him to back up his claim? If you understand what getting an education is all about, you'll choose option b. Most students would. But for some reason, those same students will dutifully scribble down every formula and example a math professor puts on the board without ever questioning where it all comes from or why it makes sense. Today, we fight back! In this section we'll study two different types of reasoning used in math. Ultimately, what it's about is being able to think in a logical, orderly way. And what skill could possibly be more useful than that?

0. After reading the opening paragraph, what do you think the main topic of this section will be?

2-6 Group

Each day that the United States Supreme Court is in session, the nine justices perform the traditional "conference handshake" that began during the late 19th century under Chief Justice Melville W. Fuller. Each justice shakes hands with all of the others to indicate that their differences of opinion won't prevent them from focusing on justice and a fair judicial process.

Here's a question to ponder: just how many total handshakes are necessary? Will there be any time left over for things like doing their actual job? It's not such an easy question. So let's see if we can look at some simpler cases to give us some guidance.

1. How many handshakes would there be if there were just two justices?

We can illustrate the situation by drawing two dots, one for each justice, and connecting them with a line, which represents them shaking hands with each other.

You might think that drawing the diagram for two justices was silly, but it gives us an idea of how to figure out the number of handshakes for larger groups.

2. Connect each pair of dots with a line to indicate a handshake that will take place. Then count the number of lines (handshakes) for each group size.

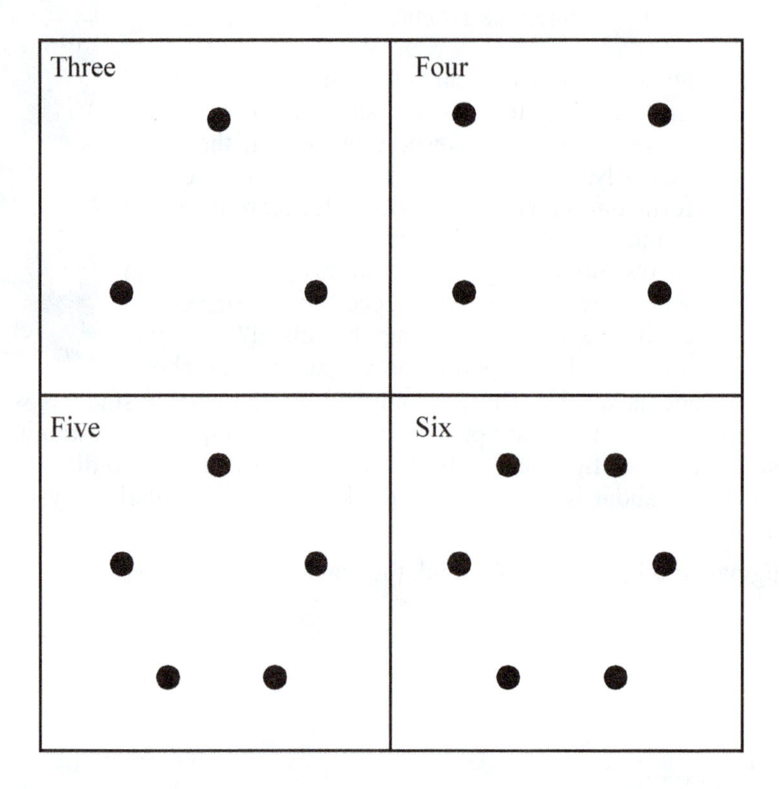

Inductive reasoning is the process of reaching a conclusion based on specific examples. After finding the number of handshakes for two through six justices, can we use those examples to draw a conclusion about how many are needed for all nine?

3. Fill in the table with the results you've already discovered, then see if you can use inductive reasoning to find a pattern and finish the rest of the table.

Number of Justices	Number of Handshakes
2	
3	
4	
5	
6	
7	
8	
9	

In Question 3, you made an educated guess about the number of handshakes necessary for nine justices using inductive reasoning. **Conjecture** is another name for an educated guess based on available evidence. Proving that a conjecture is true almost always involves more than inductive reasoning. For example, after watching a couple of races on TV, you might conjecture that all NASCAR drivers are men. In order to prove that conjecture, you would need to confirm the sex of every single driver in NASCAR; just watching a bunch of races won't prove anything conclusively.

Proving that a conjecture is false is much easier – all you need is a single example that violates the conjecture. If Danica Patrick drives in the next race you watch, she provides a **counterexample** that proves your conjecture was false.

In Questions 4–6, find a counterexample to prove that each conjecture is false.

4. Every time it rains, my softball game gets cancelled. Our game was cancelled last night.

Conjecture: It must have rained last night.

Counterexample:

5. Look at the equation $y=(x+3)(x-7)(2x-1)$.

a. Substitute –3 in for x and find the corresponding value for y.

b. Repeat part a for $x = 7$.

Conjecture: Any number substituted in for x will make $y = 0$.

Counterexample:

6. These three right triangles are all isosceles.

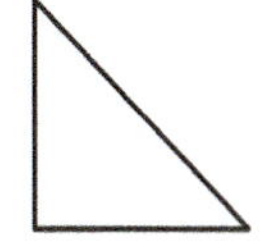

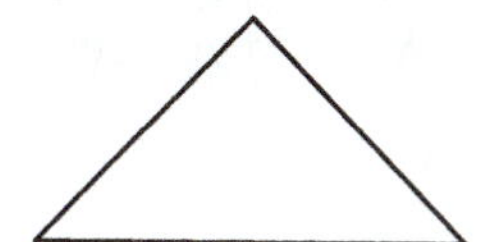

Conjecture: Every right triangle is an isosceles triangle.

Counterexample:

> **Math Note**
>
> A triangle is **isosceles** when exactly two sides have the same length.

7. The numbers illustrated below are called the **triangular numbers.** Use inductive reasoning to find the fifth and sixth triangular numbers. You can either draw diagrams or notice a pattern.

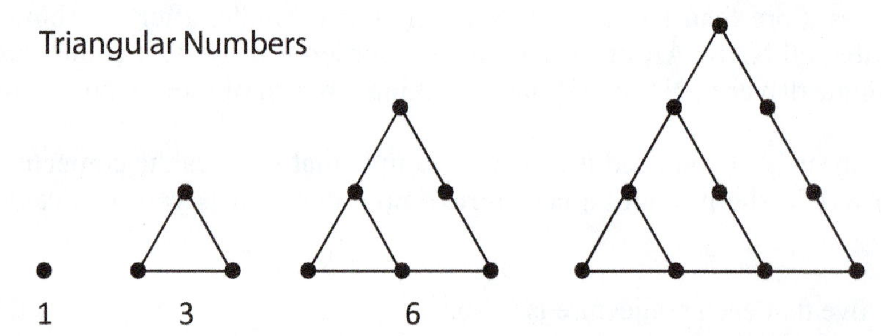

8. This pattern should look familiar .... If it doesn't, go back to Page 128. Looking at the diagrams, can you think of a reason that the triangular numbers match the number of handshakes needed, starting with two people? Or do you think it's a coincidence?

9. The triangle below is called **Pascal's triangle** in honor of the 17th-century French mathematician Blaise Pascal. Even though it's relatively simple, it turns out to have a wide variety of applications in math. Use inductive reasoning to complete the 6th and 7th rows of the triangle.

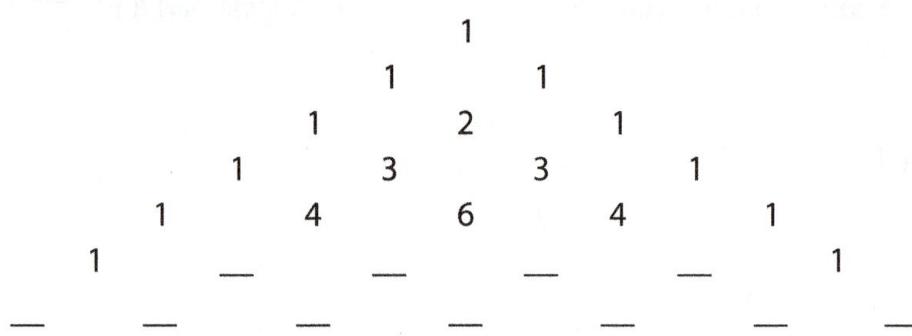

10. Here's a clever number trick. Pick any number you like, and multiply it by 12. Add 30, then divide by 6. Subtract 5 from the result, then subtract your original number. What do you get? Try some specific examples, and use inductive reasoning to make a conjecture as to what the result will always be.

2-6 Class

Now it's time to take a closer look at deductive reasoning. Inductive reasoning is great for seeing patterns and making educated guesses about general results. For example, after completing Question 10 in the Group section, you're probably pretty darn sure that the result of the number trick is always the original number. But no matter now many different numbers you choose, you can't be 100% certain that it ALWAYS works that way unless you try every number in the world. Of course, that's impossible.

This is where deductive reasoning comes into play. Instead of looking at specific examples, we'll start with an **arbitrary number** and represent it with the letter n. By arbitrary, we mean a number that is able to represent ALL possible numbers. Note the connection to our study of variables: again, we're using a letter to represent a quantity (the number we're starting with) that can change.

1. Starting with a number n, multiply that number by 12.

2. Now add 30 to the result of Question 1.

3. Next, divide the result of Question 2 by 6. Be careful! Make sure you divide BOTH TERMS by 6.

4. Subtract 5 from the result of Question 3.

5. Now subtract the original number n. What is the result?

6. Describe why this is different than the specific calculations we did in the group activity, and why it PROVES that the result will ALWAYS be the original number.

This is an example of deductive reasoning because we used a general calculation to draw a conclusion, not a handful of specific calculations. Let's see if we can use a similar approach to find a general formula that solves the handshake problem once and for all.

7. Pretend that you're one of the nine justices: how many people's hands do you need to shake?

8. Choose one other justice. How many hands does he or she need to shake?

9. If we number the justices from 1 through 9, this table shows how many hands each needs to shake.

Justice	1	2	3	4	5	6	7	8	9
Number of handshakes	8	8	8	8	8	8	8	8	8

This results in how many handshakes total? (Write as a multiplication.)

10. This result doesn't match the answer we got using inductive reasoning. Explain why this answer is twice as big as it should be. (Think about handshakes, not math!)

11. Conclusion: The number of handshakes needed for nine justices is $\frac{9\times 8}{2}$. Write a general formula with variable n that would describe the number of handshakes needed for a group of n people. Then verify that your formula gives the values you found for the table on page 128.

To wrap up our study of inductive and deductive reasoning, decide if the type of reasoning used in each example is inductive or deductive.

12. Elaine did really well on her first two exams, so she decided that she didn't need to study very much for the third exam.

13. Jake's credit card charges an insanely high interest rate, so he decided to pay off the full balance every month.

14. Chanel eats a diet very high in saturated fat, so she wasn't surprised at all when her clothes started getting tight.

15. We've won 12 consecutive games against our arch-rival, so I have no doubt we'll beat them this year.

2-6 Portfolio

Name ______________________________

Check each box when you've completed the task. Remember that your instructor will want you to turn in the portfolio pages you create.

Technology

1. ☐ Complete the spreadsheet below down to the row for 30 people. In the Group section, you noticed a pattern using inductive reasoning: use that pattern to write formulas in column B. In the Class section, we developed a formula for the number of handshakes required. Use that formula in column E. A template to help you get started can be found in the online resources for this lesson.

	A	B	C	D	E
1	Inductive Reasoning			Deductive Reasoning	
2	Number of people (n)	Number of handshakes (H)		Number of people (n)	Number of handshakes (H)
3	2	1		2	
4	3	3		3	
5	4	6		4	
6	5			5	

Skills

1. ☐ Include any written work from the online skills assignment along with any notes or questions about this lesson's content.

Applications

1. ☐ Complete the applications problems.

Reflections

Type a short answer to each question.

1. ☐ Describe the difference between inductive and deductive reasoning.
2. ☐ Describe a time in your life that you used inductive reasoning, and one when you used deductive reasoning.
3. ☐ What's the point of a counterexample?
4. ☐ Take another look at your answer to Question 0 at the beginning of this lesson. Would you change your answer now that you've completed the lesson? How would you summarize the topic of this lesson now?
5. ☐ Do you have any questions about this lesson?

Looking Ahead

1. ☐ Read the opening paragraph in Lesson 2-7 carefully, then answer Question 0 in preparation for that lesson.

2-6 Applications Name ______________________________

In the Group section, we looked at triangular numbers. We can also define square numbers and pentagonal numbers.

1. Use inductive reasoning to find the fifth square number.

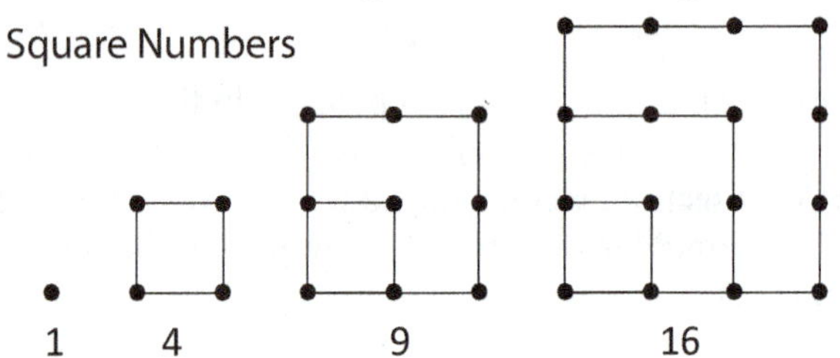

2. Use inductive reasoning to find the fifth pentagonal number.

Pentagonal Numbers

1 5

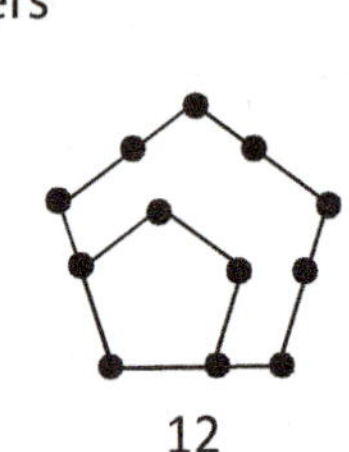

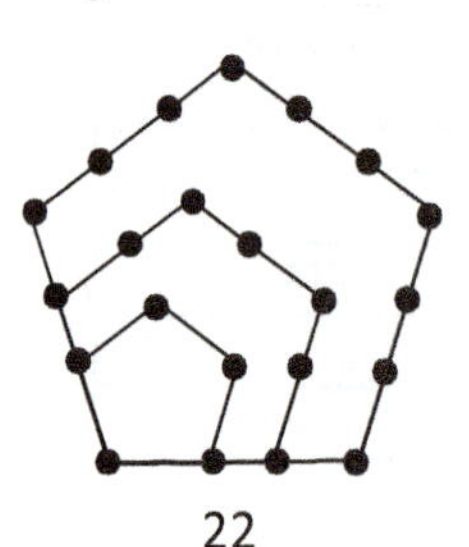

3. A formula for the nth triangular number is $\frac{n}{2}(1n-(-1))$; this simplifies to $\frac{n(n+1)}{2}$.

A formula for the nth square number is $\frac{n}{2}(2n-(0))$; this simplifies to n^2.

A formula for the nth pentagonal number is $\frac{n}{2}(3n-(1))$; this doesn't simplify. Sigh.

Using inductive reasoning, a formula for the nth hexagonal number is $\frac{n}{2}(\_\_n-(\_\_))$.

(Hint: Look at the pattern in the missing numbers for the previous three formulas.)

4. Use the formulas from Question 3 to complete this table. Feel free to use a calculator, or (better still) a spreadsheet.

	Triangular Numbers	Square Numbers	Pentagonal Numbers	Hexagonal numbers
1st				
2nd				
3rd				
4th				
5th				
6th				
7th				

2-6 Applications Name ____________________

In Questions 5 and 6, find a counterexample that shows the conjecture is false.

5. Conjecture: No day of the week has the letter h in it.

6. Conjecture: The name of every state is at least five letters long.

In Questions 7-10, decide if inductive or deductive reasoning was used.

7. The last six U.S. Representatives from this district were Republican. I don't know why the Democrats are even bothering to back a candidate this year.

8. In this state, to be a nurse you need at least a two-year degree. So when I was in the emergency room after an accident, I asked the super-cute nurse where she went to college.

9. Since it gets dark around 6 PM at this time of year, I planned my day so that I'll get home from running by 5:45.

10. Last night while playing blackjack, the dealer's first card was an ace four times in a row. So I really couldn't believe it when it happened again on the next hand.

Lesson 2-7 What's Your Problem?

Learning Objectives

- ☐ 1. Identify the four steps in Polya's problem-solving procedure.
- ☐ 2. Solve problems using a diagram.
- ☐ 3. Solve problems using trial-and-error.
- ☐ 4. Solve problems requiring calculations.

Avoid problems, and you'll never be the one who overcame them.
– Richard Bach

It just occurred to me that snakes and word problems in math have a lot in common. Most people are afraid of them, and that fear is usually irrational. In each case, education is the key to overcoming your fear. While you may be able to avoid snakes, you can't avoid problems – they're an inevitable part of life. And there's a good reason that math problems are called "problems"! The skills and thought processes you practice when working your way through word problems will come in handy when dealing with other types of problems outside of the hallowed halls of academia. So in this lesson, we'll focus on problem-solving techniques.

0. After reading the opening paragraph, what do you think the main topic of this section will be?

2-7 Class

1. Write a brief description of two or three problems that you've had to solve outside of school.

WORKING VS. DOING HOMEWORK

GETTING A RIDE TO THE DOCTOR

GETTING A RIDE TO GROCERY SHOP

2. What are some methods that you use to solve problems in your life? Do you think that they apply to solving problems in college classes?

HECK NO

A Hungarian mathematician named George Polya published a book on problem solving in 1945. In it, he described a problem-solving procedure that he felt was common to most of history's greatest thinkers. His book has been translated into at least 17 languages and is still a big seller on Amazon, so obviously he must have been on to something! Polya's procedure isn't exactly earth-shattering: its brilliance really lies in its simplicity. Polya's basic steps are listed below, followed by our spin on each step.

Polya's Four-Step Procedure For Solving Problems

Step 1: Understand the problem. The best way to start any problem is to write down relevant information as you read it. Especially with longer word problems, if you read the whole problem at once and don't DO anything, it's easy to get panicked. Instead, read the problem slowly and carefully, writing down information as it's provided. That way, you'll always at least have a start on the problem. Another essential step: carefully identify AND WRITE DOWN what it is they're asking you to find; this almost always helps you to devise a strategy.

Step 2: Devise a plan to solve the problem. Good planning is the key to any successful operation! This is where problem solving is at least as much art as science—there are many, many ways to solve problems. A short list of common strategies includes making a list of possible outcomes; drawing a diagram; trial and error; finding a similar problem that you already know how to solve; and using arithmetic, algebra, geometry, or good old-fashioned common sense.

Step 3: Carry out your plan to solve the problem. There's no point in making a plan if you don't carefully execute it. If your original plan doesn't work, don't settle for failure and give up – try a different strategy! There are many different ways to attack problems. Be persistent!

Step 4: Check your answer. In ANY problem-solving situation, you should think about whether or not your answer is reasonable. In many cases you'll be able to use math to check your answer and see if it's exactly correct. If not, don't forget what we learned about estimation in Lesson 1-4—that can be a big help in deciding if an answer is reasonable.

Now let's practice using Polya's procedure.

3. When an average-sized person walks at a fairly brisk pace, he or she can burn about 100 calories in one mile. A standard weight-loss rule of thumb is that you need to burn 3,500 calories to lose 1 pound of body fat. How many miles would you have to walk per day to lose one pound in one week?

Step 1: Understand the problem. Write down the relevant information provided in the problem.

a. How many calories are burned per mile?

b. How many calories do we need to burn in a week?

Identify what we're asked to find.

c. What exactly did the problem ask us to find?

Step 2: Devise a plan. Since we know that we need to burn 3,500 calories in a week, we can calculate how many we need to burn in one day. Then we can use that amount to figure out how many miles we'd need to walk (given that we burn 100 calories per mile).

Step 3: Carry out the plan.

d. At 3,500 calories per week, how many would we need to burn in one day?

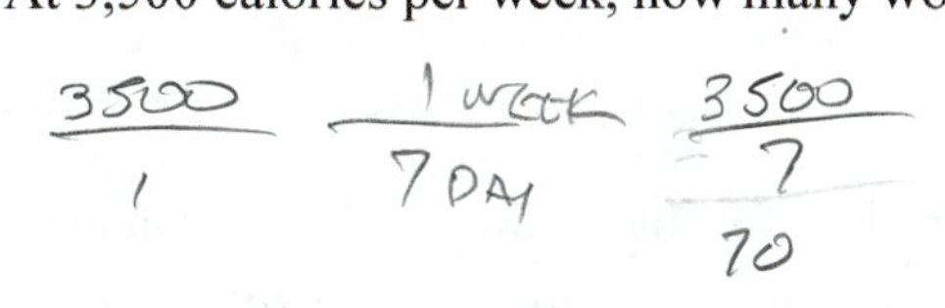

e. At 100 calories per mile, how many miles would we need to cover to burn the number of calories found in part d?

Step 4: Check the answer.

f. With the number of miles you found in part e, how many calories would we burn in one day? In one week? Have you solved the problem correctly?

A useful reminder before moving on: the two most important elements of understanding the problem are the two things you should ALWAYS do when attacking a word problem:

Key First Steps In Attacking Word Problems

1. Write down any relevant information AS YOU COME TO IT, rather than reading the whole problem all at once. If you're not sure if something is relevant or not, write it down and decide later.

2. Identify AND WRITE DOWN exactly what it is the problem is asking you to find. After you've done that, carefully read the problem again to make sure you understand the context.

2-7 Group

In this activity, you'll be working on four problems of different types. We'll help you to follow Polya's general guidelines on the first. After that, you're on your own. I know you won't let me down.

1. You have 3 picture frames you plan to hang next to one another on a section of wall 48 inches wide. The frames are each 10 inches wide. To make them look good on the wall, you plan to make the blank wall space between each frame identical, and put that same amount of space between the ends of the wall and the edge of the nearest frame. Each frame has a single hook in the center of its back. Where should you put in the three nails needed to hang the frames?

Step 1: Understand the problem.

a. Write down the relevant information provided by the problem.

3 10" PICTURE FRAMES
48" OF WALL SPACE

b. What exactly are you being asked to find?

WHERE THE NAILS GO IN THE WALL

Step 2: Devise a plan.
When a problem describes something physical that can be drawn, a diagram is usually a good idea.

c. Draw a diagram based on the description of the problem including the relevant information you wrote down.

Based on your diagram, you should be able to note how much space will be covered by frames, and how many blank spaces there will be. Once you know that, you can find the total blank space and divide by the number of spaces to find how far from the edges of the wall the edge of each frame will be.

Step 3: Carry out the plan.

d. How much of the 48 inches will be covered by the frames? How much will be left blank?

30" COVERED 18 BLANK

e. How many empty spaces are there? How wide should each empty space be? Use this information and your diagram to decide where the nails should be placed.

THERE ARE 4 EMPTY SPACES

3 × 10 FRAMES = 30
48" WALL SPACE − 30 FRAMES = 18

18 IN SPACE / 4 SPACES = 4.5 EQUAL SPACES

FROM ONE WALL THE FIRST NAIL WILL BE AT 9.5 INCHES THE SECOND NAIL WILL BE AT 24 INCHES THE 3RD NAIL IS AT 38 INCHES

Step 4: Check the answer.

f. Use your diagram to decide if the spots you found for the nails will work according to the statement of the problem.

Now you're on your own. Remember that the point of this activity is to think about a problem-solving procedure that can be used on any problem, so focus on our systematic procedure rather than just trying to figure out an answer.

2. A standard football field is 120 yards long and 160 feet wide. How long would it take Forrest Gump to mow a football field if his mower cuts a 30-inch-wide path and travels 4 miles per hour?

3. Three coworkers from Dunder Mifflin purchased a lottery ticket and won the big prize. They thought they would each get 1/3 of the winnings. As they were filing paperwork to process the winning ticket, they found out that the state and federal government would get a combined 2/5 of the winnings in taxes. What percentage of the original jackpot does each person get?

$\frac{2}{5} = 40\%$

$1 = \frac{5}{5} - \frac{2}{5}$

$40\% - 100\%$

$= \frac{60\%}{3} = 20\%$

EACH PERSON WILL GET ONE FIFTH OF THE JACK POT

4. Jack and Diane are going to start saving for retirement. They deposit $1,000 in an account to start out. Then they plan to deposit $50 each month. How long will it take until they've deposited $4,500?

Lesson 2-7 What's Your Problem

Steps for Solving Word Problems

1. Read the problem – many times – and write down the relevant information as you come to it.

2. Determine what you are asked to find – what is the question

3. Devise a plan to solve the problem – draw a picture, use a formula, do some math, make a table

4. Carry out the plan – do the work

5. Check your answer to be sure it makes sense

Solve each problem using Polya's Method

a) Roberta has 420 feet of fence and she wants to fence in her garden. If the garden is twice as long as it is wide, what is the length and width of her garden?

1. Relevant Information – 420 feet of fence
   - the length is 2 times the width – $L = 2w$
2. What do you need to find – the length and width of her garden
3. Plan – Formula – $P = 2L + 2w$
   $P = 420$ and $L = 2w$
4. Carry out the plan – $420 = 2(2w) + 2w$
   $420 = 4w + 2w$
   $\frac{420}{6} = \frac{6w}{6}$
   $70 = w$
5. Answer - The width is 70 feet and the length is twice that or 140 feet

b) A soft drink machine requires $1.25 for a can of cola. How many combinations of nickels, dimes, and quarters are possible to get exactly $1.25?

1. Relevant information – Has nickels, dimes, and quarters
   Needs $1.25
2. What do you need to find – all possible combinations that add up to $1.25
3. Plan – Make a table
4. Carry out the plan

Nickels	Dimes	Quarters	Total
0	0	5	1.25

5. Answer

25

1 12

Practice Problems for Lesson 2-7

Problem Solving Examples (show all steps described in class).

1. Roberta has 440 feet of fence. After fencing in a square region, she has 136 feet of fence left. What is the length of one side of the square region?

2. I had some pennies, nickels, dimes, and quarters in my pocket. When I reached in and pulled out some change, I had less than 10 coins whose value was 42 cents. How many possibilities are there for the coins I had in my hand?

3. Sue Ellen and Angela both have $510 in their savings accounts now. They opened their accounts on the same day, at which time Sue Ellen started with $70 more than Angela. From then on Sue Ellen added $10 to her account each week, and Angela put in $20 each week. How much money did Sue Ellen open her account with?

4. I need to build a rectangular pen for my chickens. I want the pen to have a length that is three times the width, but I only have 32 feet of fencing material to work with.
   a. What will be the dimensions of the pen I will be able to make given these restrictions?
   b. As a rule of thumb, each chicken will need 4 square feet of space inside the pen in order to live comfortably. How many chickens will I be able to keep in this pen?

2-7 Portfolio Name ____________________

Check each box when you've completed the task. Remember that your instructor will want you to turn in the portfolio pages you create.

Technology

1. ☐ Trial-and-error isn't the greatest strategy for solving every problem, but it does come in handy from time to time. And technology can help shoulder the burden of repeated calculations. Suppose that Jack and Diane from Group Question 4 start with $2,000 and invest $175 every month. Set up a spreadsheet with number of months in column A and the value of the account in column B. Use a formula that you can copy down each column after putting in the original values. Then find how long it will take for the retirement account to reach $10,000. A template to help you get started can be found in the online resources for this lesson.

Skills

1. ☐ Include any written work from the online skills assignment along with any notes or questions about this lesson's content.

Applications

1. ☐ Complete the applications problems.

Reflections

Type a short answer to each question.

1. ☐ Summarize each step of Polya's problem solving strategy in your own words.
2. ☐ Write an honest assessment of your experience with word problems in math: the good, the bad, and the ugly. What do you think is most responsible for these experiences, either good or bad?
3. ☐ Take another look at your answer to Question 0 at the beginning of this lesson. Would you change your answer now that you've completed the lesson? How would you summarize the topic of this lesson now?
4. ☐ Do you have any questions about this lesson?

Looking Ahead

1. ☐ Download and look over the article about renting a car from the online resources for Lesson 2-8. You don't have to read every word, but familiarize yourself with the argument the rental company is trying to make. The group section of the next lesson refers to the article.
2. ☐ Read the opening paragraph in Lesson 2-8 carefully, then answer Question 0 in preparation for that lesson.

2-7 Applications

Name WILLIAM BEZZER

1. Rick took his family to a major league baseball game. He gave his son $100 to buy food and drinks for the group. Hot dogs cost $3.50 each, beer costs $8.00 each, and soda costs $5.00 each. His son returned with $25.50 in change, and Rick was arrested for sending a nine-year-old to buy beer, but that's neither here nor there. There were 8 beverages total in the order. How many hot dogs, beers, and sodas were purchased?

Information
C = CASH
G = CHANGE
P = SPENT
B = $8.00
S = $5.00
H = $3.50
A = ANSWER

Find
How many hot dogs, beers & sodas?

Plan
C − G = B
$100.00 − $25.50 = $74.50

A = B(8) + H(3)
A = $64.00 + H(3)
A = $64.00 + 10.50 = $74.50

A = B(1) + S(7) + H(9)
A = 8 + S(7) + H(9)
A = 8 + 35 + H(9)
A = 8 + 35 + 31.50 = $74.50

Answer
RICK'S SON PAID FOR 8 BEERS AND 3 HOT DOGS AND NO SODAS

OR

RICK'S SON PURCHASED 1 BEER, 7 SODAS, AND 9 HOT DOGS

2. On the way to the airport on December 8, 2011, my wife's car hit a huge hole in a construction area, damaging the front right wheel. The temporary tire forced us to decrease our average speed by 15 miles per hour. If the 39-mile drive usually takes us 45 minutes, how much time should we have budgeted for the drive home from the airport? INFORMATION: 39 MILES IN 45 MINUTES, REDUCED SPEED 15 MPH

PLAN 39 MILES / 45 MINS = 0.87 / 1 MIN · 60 MIN / 1 MILE X = 52 MPH

FIND, HOW MUCH TIME WILL IT TAKE AT THE LOWER SPEED

52 MPH − 15 MPH = 37 MPH

39 MILES / 1 · 1 MILE / 37 MPH = 39/37 = 1.054 HRS

.054 × 60 MIN = 3.24

.24 × 60 SEC = 14.4

= 1 HR 3 MINS, 14 SEC

.054 HR · 60 min / 1 hr = 3.24   .24 MIN / 1 · 60 SEC / 1 MIN = 14.4

ANSWER:
WE SHOULD BUDGET ~~1 HR 3 MINS 14 SECONDS~~ FOR THE DRIVE HOME FROM THE AIRPORT

2-7 Applications Name WILLIAM BEZZER

3. A landscape architect is planning a new nature area in the middle of an urban campus. She wants the length to be twice the width, and wants to put a 3-foot high retaining wall around the perimeter. There will be 300 total feet of wall installed. How wide will this area be?

INFORMATION: USING 300 TOTAL FEET OF WALL WHERE THE LENGTH IS TWICE THE WIDTH.

FIND: THE WIDTH OF THE NEW NATURE AREA

PLAN: L 2, W 1, W 1, L 2

300 FEET FENCE ÷ 6 SECTIONS FENCE = 50 FT PER SECTION

L 2(2) + W1(2)

L = 4 + W1(2)

L = 4 + W 2

L = 4 ÷ 2 SIDES = 2 50 FT SECTIONS

W = 2 ÷ 2 SIDES = 1 50 FT SECTION

L = 4

+ W = 2

6 SECTIONS

P = 300

3W = 6W

50 = W

ANSWER

THE NEW NATURE AREA ON CAMPUS WILL BE 50 FEET WIDE

4/5

4. A pool is being built in a new student rec center at Falcon Community College. The pool is designed to be a 60 ft by 26 ft rectangle, and the deck around the pool is going to be lined with slate tiles that are 1 ft squares. How many tiles are needed? (This is not quite as easy as it seems at first...)

C = CORNERS L = LENGTH W = WIDTH

INFORMATION: A POOL THAT IS 60 FT BY 26 FT TO BE LINED BY 1 FT SQUARE TILES

FIND: THE NUMBER OF 1 FT SQUARES THAT WILL LINE THE POOL

PLAN

L = 60(2) + W = 26(2) + C why?

L = 120 + W = 26(2) + C

L = 120 + 52 + C

= 120 + 52 + 4 = 176

4/5

ANSWER

THERE ARE 176 ONE FOOT SQUARE TILES NEEDED TO LINE THE DECK AROUND THE POOL

Lesson 2-8 Indecision May or May Not Be My Problem

Learning Objectives

- ☐ 1. Consider major life decisions that involve answering "Which is better?"
- ☐ 2. Decide if one option is better by looking at values in a table.

Everyone has to make their own decisions. I still believe in that. You just have to be able to accept the consequences without complaining.
– Grace Jones

Making your own decisions is both the blessing and the curse of being an adult. While every young person strives for being able to determine their own path, there's a certain comfort in someone else making the big decisions for you. But like it or not, adulthood comes calling for all of us eventually, and we need to make decisions every day, from trivial ones like what to have for breakfast to important ones like what career to pursue. Flipping a coin is a simple way to make decisions, but obviously you won't get very far in life if all your decisions are made based on the whims of a long-dead statesman immortalized in copper. So now is a good time to make an important decision in life: decide to get better at making decisions!

0. After reading the opening paragraph, what do you think the main topic of this section will be?

2-8 Class

It seems like as you get older, more of the decisions you're faced with involve money somehow. Still, there are plenty of other considerations that go into decision making: time, convenience, compassion, and sometimes plain old fun. Here's a VERY partial list of some decisions you may be faced with:

Where to live	What degree to pursue	Buy or rent housing
Buy or lease a car	Cell phone plan	TV/Internet provider
Buffet or order from menu	Name brand or generic product	Where to shop

1. Add at least four items to the important decision list that have some significance to your life.

GO TO SLEEP OR DO HOMEWORK
EARN MONEY OR DO HOMEWORK
SLEEP OR DUST MY ROOMS
MAKE DINNER OR NOT EAT

2. Pick two or three decisions from the list and make a list of factors that you'd consider when making your decision.

NAME BRAND OR GENERIC PRODUCT — SLEEP OR DUST MY ROOMS

2-8 Group

The article you were asked to look at in preparation for this lesson is a marketing tool used by a car rental company to try and persuade potential customers that renting a car is a wise financial decision. Let's see just how wise a decision it is.

The article says that renting a full-size car for a weekend trip of 300 miles or less will cost $68.97 for the rental, plus $0.39 per mile. That $0.39 represents costs of owning your car whether you drive it or not, so that still accrues even if you're driving a rental. We can write a formula to describe the total cost, using the number of miles as input and the total cost of renting as output:

Total cost = Rental fee plus 39 cents per mile

$$C = 68.97 + 0.39m$$

where m represents the variable number of miles that you can drive. In contrast, the article claims that the total cost of driving your own car, when you factor in gas, depreciation, maintenance costs is $0.63 per mile, so the cost of driving your car for m miles is

$$C = 0.63m$$

The cost of driving a variety of different mileages are shown in the table: they were calculated using the cost formulas we developed.

.39¢ .63¢

Distance driven (mi)	Full-size rental: Total cost	Driving own car: Total cost	Full-size rental: Average cost per mile	Driving own car: Average cost per mile
0	$68.97	$0.00	N/A	N/A
50	$88.47	$31.50	$1.77	.63¢
100	$107.97	$63.00	$1.08	.63¢
150	$127.47	$94.50	.85¢	.63¢
200	$146.97	$126.00	.73¢	.63¢
250	$166.47	$157.50	.67¢	.63¢
300	$185.97	$189.00	.62¢	.63¢

1. Use the table feature on a graphing calculator to verify the costs in the table. (See the technology box on page 150.)

2. Is there a number of miles for which the total costs for renting and driving your own car are the same? If so, find the number of miles. You'll have to experiment with some settings on your calculator.

YES ABOUT 287 MILES

3. Based on your answer to Question 2, explain when it would be cheaper to drive your own car, and when it would be cheaper to rent a full-sized car.

IT WOULD BE CHEAPER TO DRIVE YOUR OWN CAR UNDER APPROX. 287 MILES, IT WOULD BE CHEAPER TO RENT A FULL SIZE CAR OVER 287 MILES

4. Calculate the average costs per mile and put them in the table. The "per mile" is a hint on how to do it.

FULL SIZE RENTAL = 5.72 ÷ 6 = 0.95¢ COST PER MILE

OWN CAR = 378 ÷ 6 = 0.63¢

5. The marketing article claims that the average cost per mile for renting the full-sized car is $0.62 per mile. Is that claim always true? Never? Sometimes? Explain. NO, IT WOULD NOT BE TRUE UNDER 300 Miles

Next, the article claims that the deal becomes even better if the customer pays an additional $10 per day fee for unlimited miles, rather than being limited to the 300 mile max. Factoring in that $10 per day charge, plus 15% in extra fees, the total cost for the 3-day weekend rental in a full-sized car is $103.47. Add the $0.39 per mile you still pay in spite of not using your own car, and we can find a formula with input the number of miles (*m*) and output the total cost of renting (*C*).

6. Write the formula for the total cost of renting a full-sized car when the unlimited mileage charge is added.

$103.47 + .39(m)

7. At $0.63 per mile, the cost of driving your own car for *m* miles remains .63 M dollars.

Distance driven (mi)	Full-size rental: Total cost	Driving own car: Total cost	Full-size rental: Average cost per mile	Driving own car: Average cost per mile
0	$103.47	$0.00	N/A	N/A
100	$142.47	$63	1.42	.63
200	$181.47	$176	.91	.63
300	$220.47	$189	.73	.63
400	$250.47	$252	.65	.63
500	$259.47	$315	.60	.63
600	$337.47	$378	.56	.63
700	$376.47	$441	.54	.63

8. Find all the total costs and write them in the table using the table feature on a graphing calculator.

9. Is there a number of miles for which the total costs for renting and driving your own car are the same? If so, find the number of miles. You'll have to experiment with some settings on your calculator.

YES, AROUND 435 MILES

10. Based on your answer to Question 9, explain when it would be cheaper to drive your own car, and when it would be cheaper to rent a full-sized car.

IT WOULD BE CHEAPER TO DRIVE YOUR OWN CAR UNDER 435 MILES, IT WOULD BE CHEAPER TO RENT A FULL SIZE CAR OVER 435 MILES

11. Calculate the average costs per mile and put them in the table. The "per mile" is still a hint on how to do it.

.77¢ PER MILE

12. The marketing article claims that the average cost per mile for renting the full-sized car with the unlimited mileage is \$0.54 per mile. Is that claim always true? Never? Sometimes? Explain.

NO YOU WOULD HAVE TO PROVE OVER 200 MILES

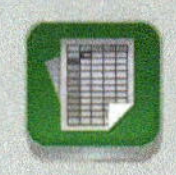

Using Technology: Making a Table on a Graphing Calculator

To make a table of values for an expression on a TI-84 Plus:

1. Press Y= to get to the equation editor screen.
2. Enter the formula you want to find values for, using X,T,θ,n for the variable.
3. Press 2nd WINDOW to get to the table setup screen. Enter the first input value you want an output for next to TblStart, and the distance between input values next to ΔTbl. (In Question 1 of the group activity, these would be 0 and 50 respectively.)
4. Press 2nd GRAPH to display the table of values.

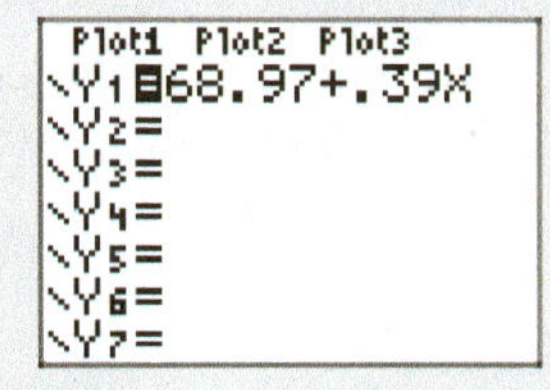

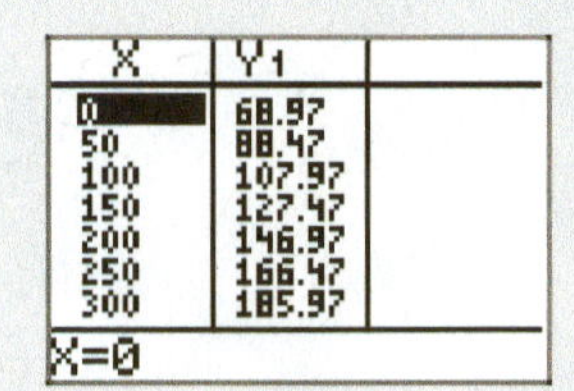

See the Lesson 2-8 video in class resources for further instruction.

2-8 Portfolio

Name ______________________________

Check each box when you've completed the task. Remember that your instructor will want you to turn in the portfolio pages you create.

Technology

1. □ In the group activity, we used the table feature on a graphing calculator to compute total costs. This could also be done with a spreadsheet, as can the average costs per mile. Build a spreadsheet that computes these values for a compact car both with and without unlimited mileage. A template to help you get started can be found in the online resources for this lesson.
2. □ Using trial and error, enter various numbers of miles for both rental options (300 miles max and unlimited mileage) until you find the number of miles that makes the rental cost as close as possible to driving your own car.

Skills

1. □ Include any written work from the online skills assignment along with any notes or questions about this lesson's content.

Applications

1. □ Complete the applications problems.

Reflections

Type a short answer to each question.

1. □ Why did you choose the life decisions you did in Question 2 of the class activity?
2. □ How does the work we did in analyzing the rental car article connect to making decisions in your life? What do you think the point of this lesson is?
3. □ Take another look at your answer to Question 0 at the beginning of this lesson. Would you change your answer now that you've completed the lesson? How would you summarize the topic of this lesson now?
4. □ What questions do you have about this lesson?

Looking Ahead

1. □ Read the opening paragraph in Lesson 2-9 carefully, then answer Question 0 in preparation for that lesson.

2-8 Applications Name ____________________

The next table shows the value remaining on a \$20 phone card based on the number of minutes the card has been used for calls. Obviously, the value remaining on a new \$20 card is \$20; from the table, we can see that the value then goes down by \$0.07 for each minute used.

Talk time (min)	Value remaining (\$)
0	\$20.00
10	\$19.30
20	\$18.60
30	\$17.90
40	\$17.20
50	\$16.50
60	\$15.80
70	\$15.10
80	\$14.40
90	\$13.70
100	\$13.00
110	\$12.30
120	\$11.60

1. The value remaining on the card can be described by the formula

20 − .07X where x is the independent variable, and represents the number of minutes used.

2. How much value remains after 50 minutes of talk time?

YOU WOULD HAVE \$16.50 REMAINING ON YOUR CARD AFTER 50 MINS OF TALK TIME

3. How many minutes can the user talk until the value remaining is down to \$13.00?

THE USER CAN TALK FOR 100 MIN BEFORE THE VALUE REMAINS \$13.00

4. Describe when the value remaining will be more than \$13.00.

ANY TIME BETWEEN 0 AND 99 MINS

5. Describe when the value remaining will be less than \$15.10.

THE VALUE REMAINING WILL BE LESS THAN \$15.10 FROM 71 MINUTES USED AND BEYOND

The two tables below display the charges for two taxi services based on the number of miles driven. Yellowish Taxi has an initial charge of \$2.30 plus \$0.15 for each quarter mile; Calloway Cab has an initial charge of \$2.00 plus \$0.20 for each quarter mile.

Yellowish Taxi

Distance (mi)	Cost (\$)
0.0	\$2.30
0.5	\$2.60
1.0	\$2.90
1.5	\$3.20
2.0	\$3.50
2.5	\$3.80
3.0	\$4.10

Calloway Cab

Distance (mi)	Cost (\$)
0.0	\$2.00
0.5	\$2.40
1.0	\$2.80
1.5	\$3.20
2.0	\$3.60
2.5	\$4.00
3.0	\$4.40

6. The cost to ride with Yellowish Taxi for m miles is \$2.30 + .60(M) dollars.

7. The cost to ride with Calloway Cab for m miles is \$2.00 + .80(M) dollars.

8. Use your answers to Questions 6 and 7 to fill in the table.

9. When would the two cabs cost the same?

THE TWO CABS WOULD COST THE SAME AT 1.5 MILES

10. When would you choose Yellowish Taxi? When would you choose Calloway Cab?

I WOULD CHOOSE CALLOWAY CAB 1.5 MILES OR LESS

I WOULD CHOOSE YELLOWISH CAB BEYOND 1.5 MILES

Lesson 2-9 All Things Being Equal

Learning Objectives

- ☐ 1. Understand and explain the meaning of solving equations and inequalities.
- ☐ 2. Develop and use procedures for solving basic equations and inequalities.

Thinking is like loving and dying. Each of us must do it for himself.

Josiah Royce, American philosopher

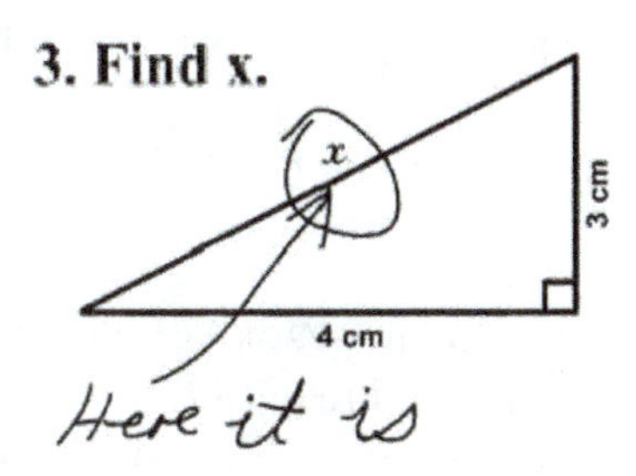

What is an equation? If your answer is "a thing you solve to find *x*," you're (a) not alone, and (b) not correct. Of all the things that contribute to students struggling in algebra, probably the biggest is a simple lack of understanding of what the important words and procedures actually mean. As the famous graphic reproduced here illustrates, not understanding what the words mean can lead to some embarrassing answers. (By the way, if you don't think that's funny, you REALLY need this lesson.) In it, we'll review basic equation-solving techniques that you probably feel like you know. In fact, you might even feel like you're expert at them. Try not to focus so much on the *how* – this lesson is about understanding the *what* and *why* of solving equations.

0. After reading the opening paragraph, what do you think the main topic of this section will be?

2-9 Group

Consider the following four statements:

- New York City is the capital of Guatemala.
- Over a million distinct species of animals have been cataloged and named.
- 8 = 5
- 13 – 10 = 3

The first and third statements are false, while the second and fourth are true. As you probably noticed, the connection between the last two statements is that each states that two things are equal. This simple exercise illustrates the two key ideas behind understanding equations.

Key Equation Ideas

1. An **equation** is simply a statement that two quantities are equal.

2. Like verbal statements, equations can be either true or false.

When an equation contains only numbers, like the two examples in our list of statements above, it's always either true or false; end of story. But many equations of interest contain variable quantities, like this one:

The number of hours you'll spend on homework this week is 14.

The number of hours you spend on homework is variable, meaning (of course) that it can change. So that statement can be either true or false, depending on what value we choose for the variable quantity. An equation like this is called a **conditional equation,** and is the type we use most commonly in areas of math (and other fields of study) that use algebra.

If we use the letter h to represent the number of hours you spend on homework, we can write our equation as $h = 14$. In this case, it should be pretty clear that if the value of h is 14, the equation is true, and otherwise it's false. So *14 is the only number that makes the equation true*, which brings us to some important definitions.

Key Equation Definitions

1. A number that makes an equation a true statement when substituted in for the variable is called a **solution** of an equation.

2. The set of ALL numbers that make an equation true is called **THE solution** of an equation.

When an equation looks like $h = 14$, the solution is pretty obvious. That's good. But when an equation looks like $-12y + 17 = 2y - 11$ ... not so much. Our goal in solving equations is to develop a systematic approach to turning an equation whose solution isn't obvious into one with an obvious solution. But that only helps if we keep this next key idea in mind, which in some sense is the most important thing in this lesson:

The Process of Solving Equations

When changing the form of an equation in an attempt to find the solution, anything you do to the equation should not affect what makes the equation true or false.

Here's a simple example of what that means: the equation $5 = 5$ is obviously true, while the equation $5 = 10$ is false. If we multiply both sides of each equation by 2:

$2 \cdot 5 = 2 \cdot 5$	$2 \cdot 5 = 2 \cdot 10$
$10 = 10$	$10 = 20$
Still true	Still false

So multiplying both sides of the equation by the same number (2 in this case) doesn't change when the equation is true and false, which means we can do that to a conditional equation *without changing the solution.* And ultimately, THAT is what this lesson is about.

1. We know the equation 1 ft = 12 in. is a true statement. So it's also true that 3 ft = 36 in. Explain why.

THERE IS 12 INCHES IN 1 FT, THEN 12 INCH X 3 FT = 36 IN

2. The equation 1 mi = 5,280 feet is true, while the equation 1 yd = 1 m is false. Multiply both sides of each equation by zero. Explain why multiplying both sides of a conditional equation by zero is a bad idea when trying to find the solution.

MULTIPLYING ANYTHING BY ZERO IS A BAD IDEA BECAUSE IT DOES NOTHING

Math Note

When two equations have the same solution, we call them **equivalent equations.** So our goal in solving equations is to transform the original equation into an equivalent equation with an obvious solution.

3. Fill in the blank to complete a description of our first useful tool for solving equations.

Equation Solving Tool #1

We can MULTIPLYING both sides of an equation by the same number or expression as long as that number or expression isn't equal to zero.

4. In Group Question 5 of Lesson 1-1, we were asked to find the number of degrees that make up a sector in a pie chart representing 35% of respondents to a poll. If we use letter d to represent that number of degrees, then the equation $\frac{d}{360} = 0.35$ can be used to solve this problem. Solve this equation using Tool #1.

$\frac{360°}{1} \cdot \frac{d}{360°} = .35 \cdot \frac{360}{1} = 126°$

D = 126°

126 = 126

5. Check that your solution is correct by showing that it makes the original equation a true statement.

$\frac{126°}{360°} = 0.35$

6. Is this the only solution to the equation? How can you tell?

YES, BECAUSE IT WORKED OUT BY CHECKING OUR SOLUTION

7. We know that the equation 1 min = 60 sec is true. So is it also true that 1/2 min = 30 sec? What did we do to both sides to produce an equivalent equation?

MULTIPLY

8. Use the result of Question 7 to fill in the blank, completing a description of our next tool for solving equations.

Equation Solving Tool #2

We can DIVIDE both sides of an equation by the same number or expression as long as that number or expression isn't equal to zero.

9. In Group Question 11 of Lesson 1-6, we were asked to find the number of fans that would sit in each row of a set of bleachers with 5 rows if 60 tickets were allotted. We could model this situation with the equation $5x = 60$, where x is the number of fans in each row. Use Tool #2 to solve this equation.

$5x = 60$

$\frac{5x}{5} = \frac{60}{5}$ $x = \frac{60}{5}$ $x = 12$

10. Show a check of your solution, and write a sentence explaining what the solution means.

5·12 = 60

IN MEANS THAT YOU WILL SEAT 12 PEOPLE PER ROW

An equation stating that two ratios are equal is called a **proportion.** The equation we solved in Question 4 can be written as a proportion if we write the percentage (35%) in fraction rather than decimal form:

$$\frac{d}{360} = \frac{35}{100}$$

A procedure called "cross multiplying" is often used to solve proportions: in this case, you'd multiply down one diagonal, giving you $100 \cdot d$, and multiply up the other diagonal, giving you $360 \cdot 35$. (This is shown in the diagram.) Setting those two results equal to each other, the resulting equation is $100d = 360 \cdot 35$, which you can solve using Tool #2. But why does this work?

$$\frac{d}{360} \times \frac{35}{100}$$

11. Use Tool #1 to multiply both sides of the original proportion by the number $100 \cdot 360$, and do any obvious reducing of fractions. What do you notice about the result?

$\frac{d}{360} \times \frac{35}{100}$

$100d = 360 \cdot 35 = 12{,}600$

$\frac{100d}{100} = \frac{12{,}600}{100}$

$d = 126$

$d = 126$

12. Use cross multiplying to solve the proportion $\frac{x}{12} = \frac{15}{36}$.

$\frac{x}{12} \times \frac{15}{36}$ $36x = 12 \cdot 15 = 180$

$\frac{36x}{36} \quad \frac{180}{36}$

$x = 5$

$X = 5$

13. We know that 1 min = 60 sec. Is it okay to add the number 3 to both sides of the equation, resulting in 4 min = 63 sec? Why or why not?

NOT, THEY ARE NOT THE SAME VALUES

14. Is it okay to add 3 seconds to both sides of the equation, resulting in 1 min 3 sec = 63 sec? Why or why not?

NOT THEY ARE STILL NOT THE SAME VALUES

15. Use the result of Question 14 to fill in the blank, completing a description of our next tool for solving equations.

Equation Solving Tool #3

We can ___ADD___ the same quantity or expression to both sides of an equation.

16. In Group Question 10 of Lesson 1-3, we found the number of meals with less than 12g of fat served at some campus cafeterias. There were 70 with less than 12 g of fat and more than 350 calories, and 142 that had less than 12g of fat and less than 350 calories. We can model this situation with the equation $x - 142 = 70$. Use Tool #3 to solve the equation.

$x - 142 = 70$
$+142 \quad +142$
$x = 212$

$x = 212$

17. Check your solution, then write a sentence describing what the answer means. What variable quantity does the letter x represent?

18. Given the fact that 36 in. = 3 ft, subtract 5 in. from each side of the equation. Is the resulting equation still true?

19. Use the result of Question 18 to fill in the blank, completing a description of our last tool for solving equations.

Equation Solving Tool #4

We can ___SUBTRACT___ the same quantity or expression from both sides of an equation.

20. In Group Question 4 of Lesson 1-2, we found that out of 2,094 respondents to a survey, 1,361 would most like to have a burger with one of the top four choices among the former presidents. We were asked to find how many chose a president who is not in the top four. We could model this question with the equation $n + 1{,}361 = 2{,}094$. Solve the equation using Tool #4.

21. Show a check of your solution, and write a sentence explaining what the solution means.

In the Group portion of Lesson 2-8, you were asked to find the number of miles driven that would make the cost of driving a rental car the same as the cost of driving your own car. We modeled the cost of driving the rental car m miles by $68.97 + 0.39m$, and the cost of driving your own by $0.63m$. We then used a table and calculator to estimate the answer to the problem.

22. Write an equation with variable m whose solution tells you the number of miles for which the two costs are equal.

23. Solve the equation. Did you get the same answer as in Question 2 of Lesson 2-8 Group?

24. Which method seems easier for finding the number of miles? Which do you think gave a more accurate answer?

In the rental car problem, it's probably more realistic to find the number of miles for which the rental car costs LESS than driving you own, as opposed to finding the number of miles for which the costs are equal. A statement that one quantity is more or less than another is called an **inequality.** The inequality $2 < 5$ is the true statement that two is less than five. The inequality $x > 3$ is a statement that the variable quantity represented by x is more than 3. Just like with equations, **the solution to an inequality is the set of all numbers that make the inequality a true statement when substituted in for the variable.**

25. Name a number that makes $x > 3$ true.

26. Name a number that makes $x > 3$ false.

Now let's develop tools to solve inequalities. First, here's a summary of the tools we can use to solve equations:

#1. Multiply the same nonzero number or expression on both sides.
#2. Divide both sides by the same nonzero number or expression.
#3. Add the same number or expression to both sides.
#4. Subtract the same number or expression from both sides.

Do these tools work for solving inequalities too? Let's see. (Remember, by "work," we mean that we can do these procedures to an inequality without changing whether it's true or false.) We'll use less than (<) to explore new rules, but the rules will also apply to greater than (>), less than or equal to ($\leq$), and greater than or equal to ($\geq$). In Questions 27-30, fill in the comparison symbol that makes the inequality true after performing the operation.

27. Add 8 to both sides.

Left side	Comparison symbol	Right
10	<	20

Left side	Comparison symbol	Right
18	<	28

28. Add –8 to both sides.

Left side	Comparison symbol	Right
10+-8	<	20+-8

Left side	Comparison symbol	Right
2	<	12

29. Subtract 6 from both sides.

Left side	Comparison symbol	Right
10 -6	<	20 -6

Left side	Comparison symbol	Right
=4	<	=14

30. Subtract –6 from both sides. (Be careful!)

Left side	Comparison symbol	Right
10-6	<	20-6

Left side	Comparison symbol	Right
16	<	26

31. Fill in either "does" or "does not" to complete our first tool for solving inequalities.

Inequality Solving Tool #1

Adding or subtracting the same number on both sides of an inequality _______________ change the direction of the comparison symbol.

32. Multiply both sides by 2.

Left side	Comparison symbol	Right
10	<	20

Left side	Comparison symbol	Right
20	<	40

33. Multiply both sides by –2.

Left side	Comparison symbol	Right
10	<	20

Left side	Comparison symbol	Right
-20	>	-40

34. Divide both sides by 5.

Left side	Comparison symbol	Right
10	<	20

Left side	Comparison symbol	Right
2	<	4

35. Divide both sides by –5.

Left side	Comparison symbol	Right
10	<	20

Left side	Comparison symbol	Right
-2	>	-4

36. Write "stays the same" or "changes" in each blank to complete our second tool for solving inequalities.

Inequality Solving Tool #2

If you multiply or divide both sides of an inequality by a positive number, the direction of the comparison symbol ____________________. If you multiply or divide both sides by a negative number, the direction of the comparison symbol ________________.

37. In Question 10 of Lesson 2-8 Group, you were asked to find when renting a full-sized car would be cheaper than driving your own. The inequality $103.47 + 0.39m < 0.63m$ can be used to model the inequality. Solve the problem. Don't forget to write your answer in sentence form.

Math Note

The inequality $3 > x$ says exactly the same thing as $x < 3$. In each case, the quantity that's less is on the closed end of the comparison symbol.

38. What does m represent in the inequality? What does $103.47 + 0.39m$ represent? What about $0.63m$?

39. Based on your answer to Question 37, should $m = 400$ make the inequality true? Explain, then substitute 400 for m to check your answer.

40. Repeat Question 39 for $m = 450$.

2-9 Portfolio

Name ______________________________

Check each box when you've completed the task. Remember that your instructor will want you to turn in the portfolio pages you create.

Technology

1. □ By encouraging you to understand what solving an equation actually means, and by asking you to check solutions by substituting back into the equation, we're hoping to help you remember that you can always check your answer when asked to solve an equation. Excel can help in that regard as well: you can build a spreadsheet like the sample below, using a formula for each of the right and left sides of the equation. When you input a variable value, you should be able to see if the two sides of the equation are in fact equal. Build an equation-checking spreadsheet for each of the equations we solved in this lesson and use it to double-check all of your solutions.

	A	B	C
1	Variable	Left Side	Right Side
2			

Skills

1. □ Include any written work from the online skills assignment along with any notes or questions about this lesson's content.

Applications

1. □ Complete the applications problems.

Reflections

Type a short answer to each question.

1. □ Explain in your own words what it means to solve an equation. If your answer doesn't use the word "true," it probably stinks.
2. □ What exactly is an equation? How does an equation differ from an inequality?
3. □ Take another look at your answer to Question 0 at the beginning of this lesson. Would you change your answer now that you've completed the lesson? How would you summarize the topic of this lesson now?
4. □ What questions do you have about this lesson?

Looking Ahead

1. □ Read the opening paragraph in Lesson 3-1 carefully, then answer Question 0 in preparation for that lesson.

2-9 Applications

Name WILLIAM BERGER

In many of the lessons so far, we've used the ideas behind solving equations or inequalities without actually using the symbols and procedures. In this assignment, you'll be asked to check answers to earlier problems by solving an equation or inequality. You'll need to locate that problem in its original location to either set up the equation, or verify your result. Each equation/inequality is given with a reference to the problem it helps to solve.

From Lesson 1-1 Class Question 1; fill in the blank with the number in the "Homework" row of the table.

$168P =$ 32

1. Solve the equation, showing your work and checking your solution.

$168P = 32$

$168P \div 168$

$P = 0.1905$

$168 \cdot 0.1905 = 32$

$32 = 32$

2. Describe what the variable in the equation represents, and what each side of the equation represents.

THE VARIABLE REPRESENTS PERSENTAGE OF

32 REPRESENT THE HOMEWORK HOURS

168 REPRESENT THE TOTAL NUMBERS SPENT ON ACTIVITIES

From Lesson 1-1 Class Question 4; fill in the blank with the number in the "Sleep" row of the table.

$.32 = \dfrac{d}{360}$

3. Solve the equation, showing your work and checking your solution.

$360 \cdot .32 = \frac{D}{360} \cdot 360$

$360 \times .32 = 115.20 = D$

$115.20 \div 360 = .32$

$.32 = .32$

4. Describe what the variable in the equation represents, and what each side of the equation represents.

THE VARIBLE REPRESENTS THE DEGREES PORTION OF THE EQUATION

.32 REPRESENT THE PRESENTAGE OF SLEEP HOURS IN A WEEK

360 REPRESENTS THE TOTAL NUMBERS OF DEGREES

2-9 Applications

Name WILLIAM BERGER

From Lesson 2-3 Applications Question 4:

$$\frac{22.09}{1{,}000} = \frac{x}{18{,}000}$$

5. Solve the equation, showing your work and checking your solution.

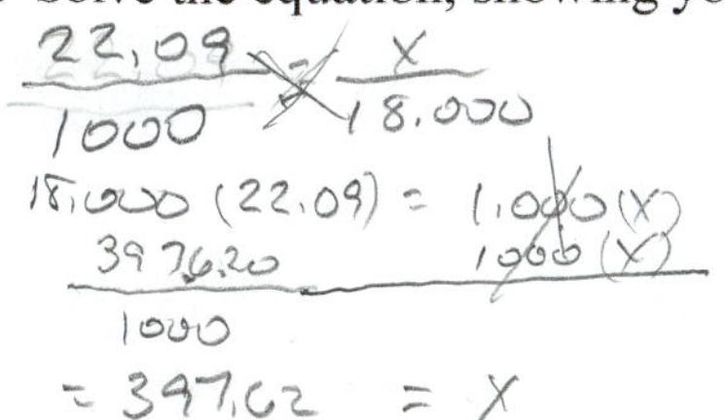

18,000 (22.09) = 1,000 (x)

397,620 / 1000 = 1000 (x) / 1000

= 397.62 = x

22.09 / 1000 = 397.62 / 18,000

.02209 = .02209

6. What does the variable represent? What is the significance of the ratio $\frac{22.09}{1{,}000}$?

THE VARIBLE REPRESENTS THE MONTHLY PAYMENT FOR EVERY $1000.00 BARROWED

IT GIVES YOU A EQUATION FOR CALCU

From Lesson 2-7 Group Question 4:

$$1{,}000 + 50x = 4{,}500$$

7. Solve the equation, showing your work and checking your solution.

1,000 + 50x = 4,500
−1,000 −1,000
0 50x 3,500 / 50
x = 70

1000 + 50(70) = 4500
1000 + 3500
4500 = 4500

8. Explain the significance of this solution, including a description of the problem that it solves.

THE SIGNIFICANCE OF THIS SOLUTION IS THAT IT SHOWS THE AMOUNT OF MONTHS IT WILL TAKE TO SAVE FOR THE VACATION

From Lesson 2-7 Application Question 3:

$$w + 2w + w + 2w = 300$$

9. Solve the equation, showing your work and checking your solution.

1w + 2w + 1w + 2w = 300
= 6w = 300
6 6
w = 50

(50) + 2(50) + (50) + 2(50) = 300
50 + 100 + 50 + 100 = 300
300 = 300

10. Explain why the variable *w* appears 4 separate times in the equation. What does *w* represent?

BECAUSE THERE ARE 4 DIFFERENT SPACES

W REPRESENTS THE WIDTH OF THE 4 SPACES

W REPRESEMS 56

From Lesson 2-8 Applications Question 10:

$2.30 + 0.60x < 2.00 + 0.80x$

11. Solve the inequality, showing all work.

$2.30 + 0.60x < 2.00 + 0.80x$

$-0.60x \qquad -0.60x$

$2.30 < 2.00 + 0.20x$

$-2.00 \qquad -2.00$

$0.30 < 0.20x$

$\frac{0.30}{0.20} < \frac{0.20x}{0.20x}$

$1.5 < x$

$= x > 1.5$

12. Describe what each side of the inequality represents.

THE DIFFERANCES IN TAKING 2 DIFFERANT CAB COMPANIES

13. Explain why an inequality makes much more sense for solving this problem than an equation.

IT MAKES MORE SENSE TO USE AN EQUALITY VERSUS A EQUATION BECAUSE IS EASIER TO CALCULATE WHEN THERE IS UNEQUE VALUES

Unit 3
Thinking Linearly

Outline

Lesson 3-1 A Coordinated Effort

Learning Objectives

- ☐ 1. Use a rectangular coordinate system.
- ☐ 2. Connect data to graphs.
- ☐ 3. Interpret graphs.

Believe you can and you're halfway there.
– Theodore Roosevelt

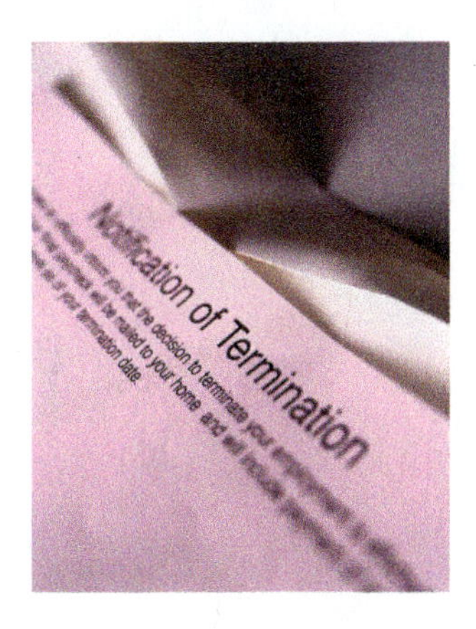

We've already used the word "graph" in this book, when referring to bar graphs. The bar graphs we looked at weren't just pretty pictures, or bars drawn at random heights for no particular reason: they were used to illustrate and understand real data. But the types of graphs we'll study in this unit might be the most misunderstood feature of math: people tend to think of them as just "plotting points and connecting the dots." Nothing could be further from the truth! Just like bar graphs, the graphs that illustrate connections between two variables are all about a visual representation of useful information. So before we talk about the mechanics and terminology involved, we'll use unemployment numbers to vividly illustrate what it's really all about.

0. After reading the opening paragraph, what do you think the main topic of this lesson will be?

3-1 Class

If you've been paying attention to the news at all for the last several years, you know that the unemployment rate in the United States has been a big story. But what did unemployment look like before the economic crisis of the late 2000s? These tables display the average annual unemployment rate for the years from 1992 to 2011.

Year	'92	'93	'94	'95	'96	'97	'98	'99	'00	'01
Rate(%)	7.5	6.9	6.1	5.6	5.4	4.9	4.5	4.2	4.0	4.7

Year	'02	'03	'04	'05	'06	'07	'08	'09	'10	'11
Rate (%)	5.8	6.0	5.5	5.1	4.6	4.6	5.8	9.3	9.6	8.9

1. Use the table to write a verbal description of trends in the unemployment rate over that 20-year period.

With enough effort, you were probably able to write a reasonable description. But because there's so much data in the table, spotting the trends isn't exactly a simple thing to do. Next, let's look at the same data displayed in graphical form.

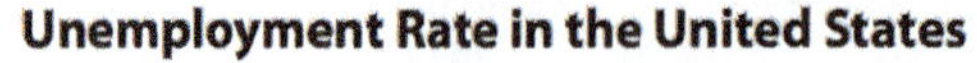

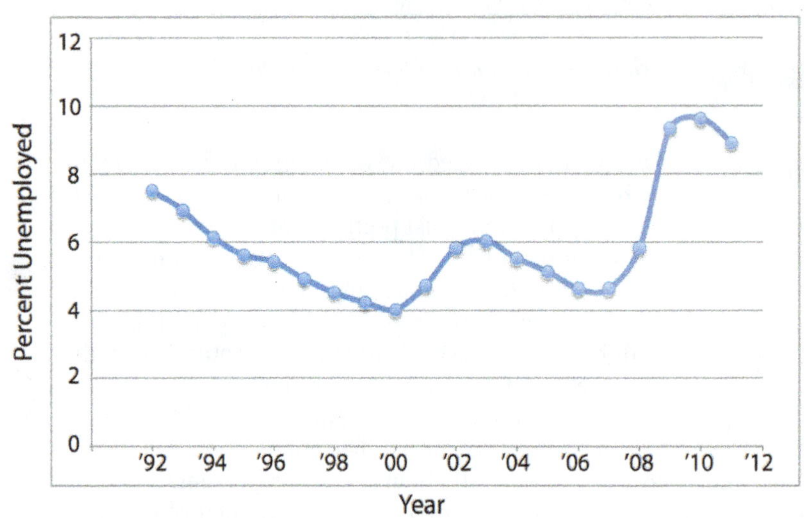

When I look at this graph, two things occur to me: it's a lot easier to see the trends than it was from looking at the table, and the graph looks a whole lot like a dinosaur, which is totally irrelevant but still pretty cool.

2. Use the graph to write a verbal description of trends in the unemployment rate, then explain why the graph makes it easier than the table did.

Without the numbers running along the bottom side of the graph and down the left side, we wouldn't be able to understand any of the information the graph provides. Those numbers provide the **scale** for the graph, and they're ALWAYS crucial in drawing a graph. Each of the number lines that we write the scale on is called an **axis** (the plural of this word is **axes**).

In preliminary numbers released in early 2013, the average unemployment rate for 2012 was listed as 8.1%. We can add that piece of information to the graph by finding 2012 on the horizontal axis and 8.1 on the vertical axis, then drawing imaginary lines up from 2012 and right from 8.1 until the lines meet: that's where we put the point corresponding to 2012 and 8.1%.

Unemployment Rate in the United States

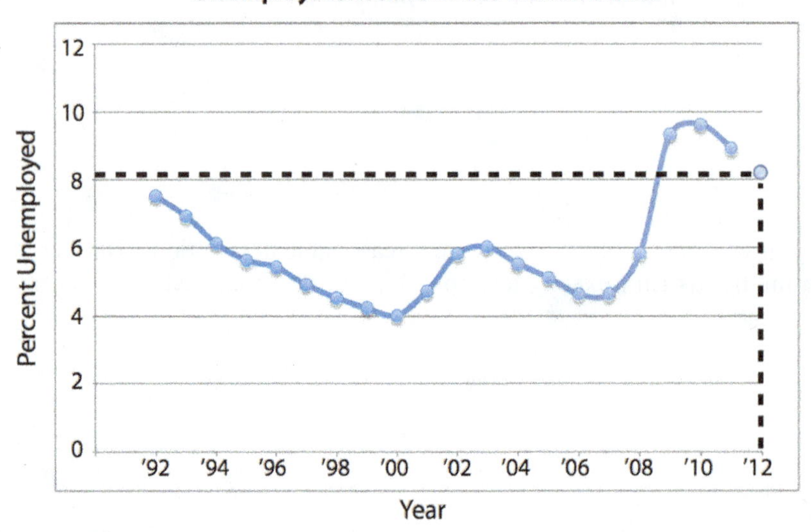

Notice that when we found the location to indicate that the unemployment rate was 8.1% in 2012, the imaginary lines we drew formed a rectangle with the two axes. That's why we call this system of graphing a **rectangular coordinate system.** Each of the numbers we used to locate that point are called **coordinates.** The horizontal axis is usually called the ***x* axis** and the vertical axis is usually called the ***y* axis.** The point where the two axes meet is called the **origin.**

Since we didn't need to worry about negative years or negative unemployment rates, the graph we drew earlier only showed positive values along each axis. But there are plenty of examples of data where negative values make perfect sense, so a rectangular coordinate system is often set up like this:

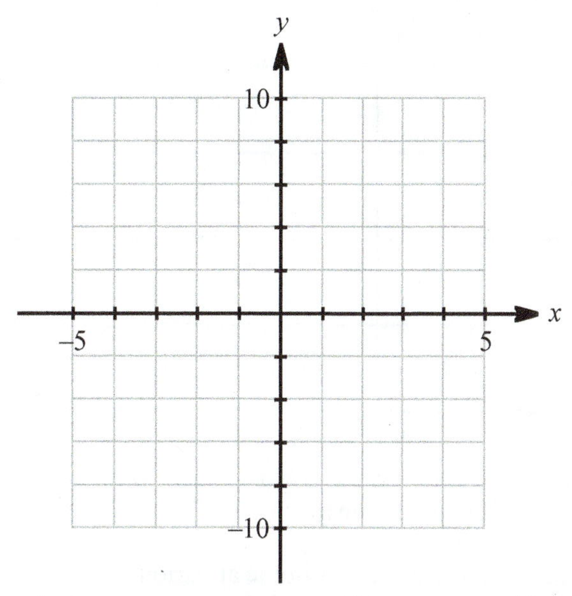

The process of locating information on a rectangular coordinate system, like the 8.1% unemployment rate in 2012, is called **plotting points.** We identify locations by writing the two coordinates together inside parentheses, like this: (2012, 8.1).

3. What are the coordinates of the origin in the rectangular coordinate system above?

4. Look carefully at the numbers on each axis. What distance does each box on the grid represent along the *x* axis? What about the *y* axis?

5. Plot each of the following points on the rectangular coordinate system above. Label the coordinates of each point.
a. (4, 0)
b. (0, –8)
c. (–3, –4)
d. (2, 7)

> **Math Note**
>
> To be more specific about the term "scale," the distances you found in Question 4 are usually referred to as the scale for each axis.

3-1 Group

If it doesn't make you uncomfortable, exchange the following information with the classmates in your Unit 3 group. This will be your small group for the third unit. It would be a good idea to schedule a time for the group to meet to go over homework, ask/answer questions, or prepare for exams. You can use this table to help schedule a mutually agreeable time.

Name	Phone Number	Email	Available times

Being able to understand the connection between a graph and the information that it illustrates is by far the most important skill in graphing. If you can't interpret the meaning of a graph, it's just really bad art!

1. A small plane takes off from a regional airport; its altitude at various times is recorded in the table below. Write ordered pairs of the form (Flight time, Altitude) for each pair of values, then plot the points on the graph.

Flight time *x* (minutes)	Altitude *y* (feet)	Ordered pair (*x*, *y*)
0	0	
10	4,000	
20	3,000	
30	2,000	
40	2,000	
50	1,000	
60	0	

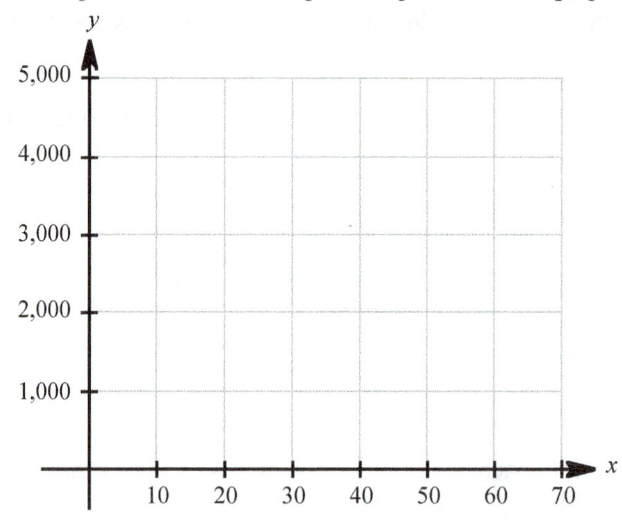

2. Connect the points you plotted to draw a graph, and describe what that graph illustrates. Then add a verbal label to each axis that describes the information it represents.

For Questions 3-8, explain how you got your answer using BOTH the table and the graph.

3. What was the highest altitude reached by the plane? (Is the answer based on your graph different? Why?)

4. How long after the flight began did the plane reach its highest altitude? (Again, you may have two slightly different answers.)

5. When was the plane flying level?

6. How long did the flight last?

7. When was the altitude of the plane increasing?

8. When was the altitude of the plane decreasing?

9. Explain why we didn't bother to include negative values along either axis when drawing the rectangular coordinate system for the altitude graph.

10. When we plot points on a coordinate system that correspond to pairs of data, we call the result a **scatter diagram** or **scatter plot.** For the bank accounts in the two tables below, create a scatter diagram for each. First, you'll need to complete the table using skills we practiced earlier in the course. After writing ordered pairs, decide on an appropriate scale for each axis, then plot each point. It would probably be a good idea to use different colors for each account. (Note: We do NOT connect the points on a scatter plot!)

Time (yrs)	Account 1	Ordered pair
Start	$1,000.00	
After 1 year	$1,060.00	
After 2 years	$1,120.00	
After 3 years	$1,180.00	
After 4 years	$1,240.00	
After 5 years		
After 6 years		
After 7 years		
After 8 years		
After 9 years		

Time (yrs)	Account 2	Ordered pair
Start	$1,000.00	
After 1 year	$1,050.00	
After 2 years	$1,102.50	
After 3 years	$1,157.63	
After 4 years	$1,215.51	
After 5 years		
After 6 years		
After 7 years		
After 8 years		
After 9 years		

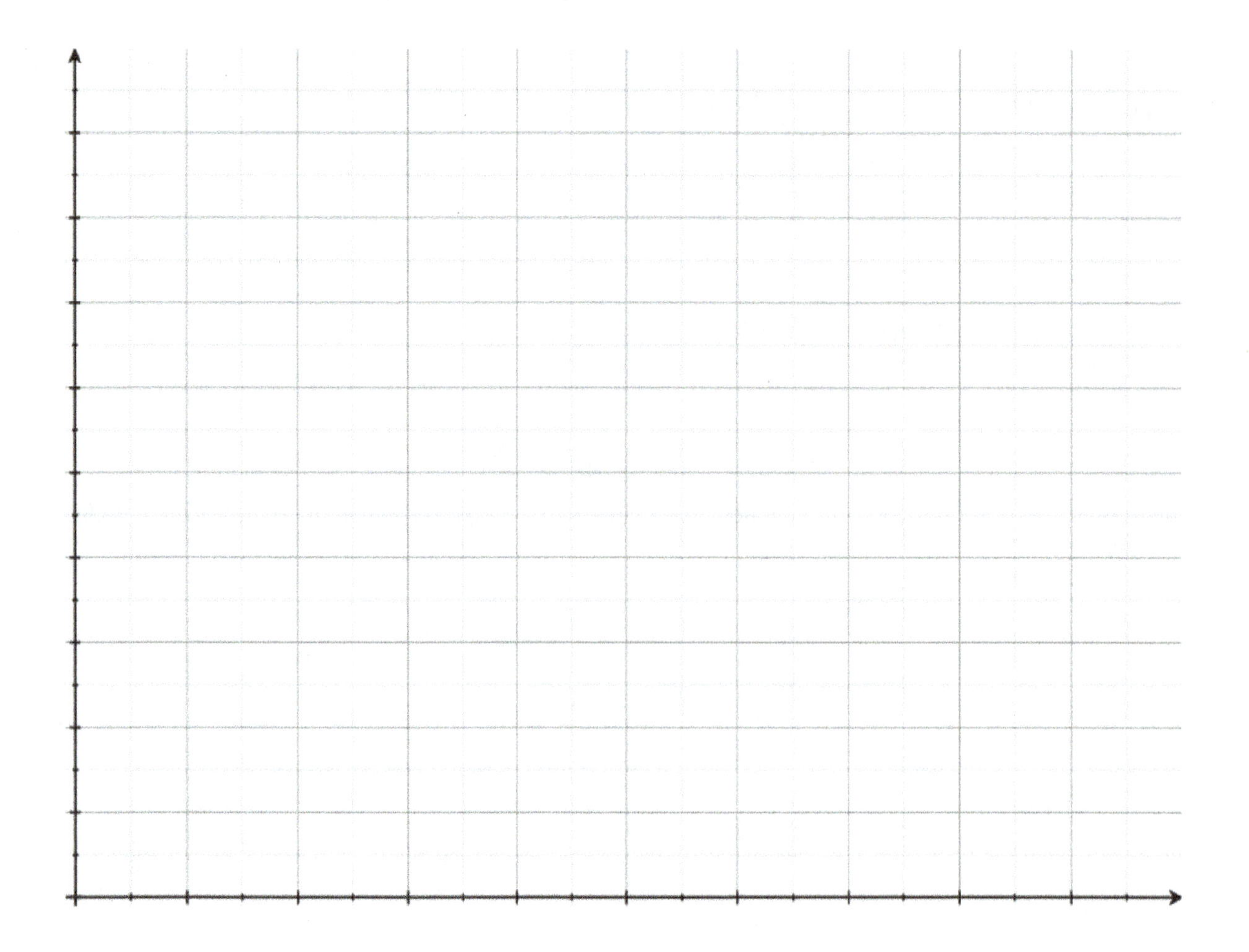

11. Explain why you chose the scale that you did for each axis.

12. Use the two scatter plots to write a verbal description of the differences between the growth of the two accounts. More detail is better!

13. In some cases, if the pattern of points in a scatter plot is relatively clear, we can connect the points to complete a graph. Do that now, drawing two graphs that represent the growth of the two accounts in Question 10. How would you describe the two graphs verbally?

Math Note

In Lesson 3-6, we'll delve deeper into when it's appropriate to connect the points on a scatter plot, and how to find a graph that best fits a given plot.

Using Technology: Creating a Scatter Diagram

To create a scatter diagram in Excel:

1. Type the values that will go on the x axis in one column. (This would correspond to the number of years in Question 10.)
2. Type the values that will go on the y axis in one column. (This would correspond to the value of the account in Question 10.)
3. Use the mouse to drag and select all the data in those two columns.
4. With the appropriate cells selected, click the **Insert** tab, then **Charts,** and click on Scatter. Then choose the type of scatter diagram you want. Options include plotting only the points, connecting the points with curves, and connecting the points with line segments.

You can add titles and change colors and other formatting elements by right-clicking on certain elements, or using the options on the **Charts** menu. Try some options and see what you can learn!

See the Lesson 3-1 video in class resources for further information.

14. What temperature is this thermometer displaying?

15. Explain why the thermometer is pretty much useless.

16. What are the coordinates of the point drawn on the graph?

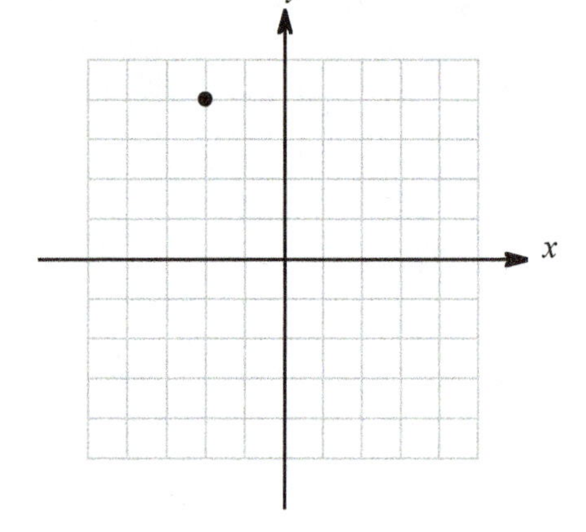

17. Explain why the graph is pretty much useless.

Important Features of a Good Graph

1. EVERY graph has to have a clearly-labeled scale on each axis. A graph with no scale labeled is every bit as useless as a thermometer with no numbers on it.

2. There's no reason the scale has to be the same on both axes. The first graph in this group activity contained points like (10, 4,000) and (20, 3,000); using the same scale on the x and y axes would lead to a graph that's very difficult to read (try it!).

3. If there are certain points on a graph that are important for some reason, you should label the coordinates of those points right on the graph.

3-1 Portfolio

Name ______________________________

Check each box when you've completed the task. Remember that your instructor will want you to turn in the portfolio pages you create.

Technology

1. □ Use Excel to create two different scatter diagrams for the ordered pairs in the table on the next page. The first should just have the points; the second should connect the points with curves. A template to help you get started can be found in the online resources for this lesson.

Skills

1. □ Include any written work from the online skills assignment along with any notes or questions about this lesson's content.

Applications

1. □ Complete the applications problems.

Reflections

Type a short answer to each question.

1. □ If someone says that the point of graphing is plotting points and connecting the dots, how would you explain to them how very, very wrong that they are? It'll be tough, but try to be nice.
2. □ Why do you think we use the word "ordered" in "ordered pair"?
3. □ Explain the advantages of graphed data over data in table form.
4. □ Take another look at your answer to Question 0 at the beginning of this lesson. Would you change your answer now that you've completed the lesson? How would you summarize the topic of the lesson now?
5. □ What questions do you have about this lesson?

Looking Ahead

1. □ Read the opening paragraph in Lesson 3-2 carefully, then answer Question 0 in preparation for that lesson.

3-1 Applications

Name ______________________________

The hourly temperatures for Champaign, Illinois on October 8, 2012 are given in the table. Use hours after midnight (NOT the actual time) as first coordinates, and the temperature as second coordinates.

Time	Temperature	Ordered Pair
12:00 AM	34°	
1:00 AM	34°	
2:00 AM	34°	
3:00 AM	36°	
4:00 AM	35°	
5:00 AM	33°	
6:00 AM	32°	
7:00 AM	33°	
8:00 AM	37°	
9:00 AM	42°	
10:00 AM	48°	
11:00 AM	51°	
12:00 PM	54°	
1:00 PM	55°	
2:00 PM	57°	
3:00 PM	57°	
4:00 PM	57°	
5:00 PM	55°	
6:00 PM	52°	
7:00 PM	49°	
8:00 PM	47°	
9:00 PM	46°	
10:00 PM	45°	
11:00 PM	44°	
12:00 AM	43°	

1. Write an ordered pair for each time and temperature pairing. Remember, the first coordinate is hours after midnight.

2. Decide on an appropriate scale for each axis and create a scatter plot for this data on graph paper. Then connect the points with a smooth curve. If you don't have graph paper, you can easily find printable graph paper online. Make sure that the lowest height on the graph corresponds to zero degrees.

3. What do you need to find in the table to find when the temperature was highest during the day?

4. What do you need to look for on the graph to find when the high temperature was reached?

5. Use your graph to estimate time spans when the temperature was increasing.

6. Use your graph to estimate time spans when the temperature was decreasing.

7. Draw a second graph for the data: this time make the lowest height on the graph correspond to 30°. Why is this second graph deceiving in terms of how much the temperature varies?

Lesson 3-2 Cabbing It

Learning Objectives

- ☐ 1. Define slope as a constant rate of change.
- ☐ 2. Define and interpret the intercepts of a line.
- ☐ 3. Tell a story based on the graph of a line.

The only thing constant in life is change.
– François de La Rochefoucauld

We were first introduced to the important idea of rates of change in Lesson 2-3. In short, a rate of change measures how some quantity is changing. Now ponder this question: do rates of change change? That might look like a typo, but it's not. The rate at which some quantity changes either stays constant, or changes. For example, if you make $9 per hour at a job, the rate at which your pay changes as you work more hours is always the same: your pay grows at the rate of $9 per hour. But if you're driving in traffic, the rate of change of your position (what we commonly call speed) probably changes quite a bit. In this lesson, we're going to study situations where the rate of change of a quantity stays constant: in that case, modeling that situation with an equation or a graph is very manageable.

0. After reading the opening paragraph, what do you think the main topic of this section will be?

3-2 Class

1. According to taxifarefinder.com, a cab ride from the airport in Las Vegas will cost you an initial charge of $5.10, plus $2.60 per mile. Use this information to fill in the rest of the table describing the total cost for various distances. Then create a scatter diagram, using the distance as x coordinate and the cost as y coordinate.

Distance	Cost ($)
0	$5.10
1	$7.70
2	$10.30
3	
4	
5	

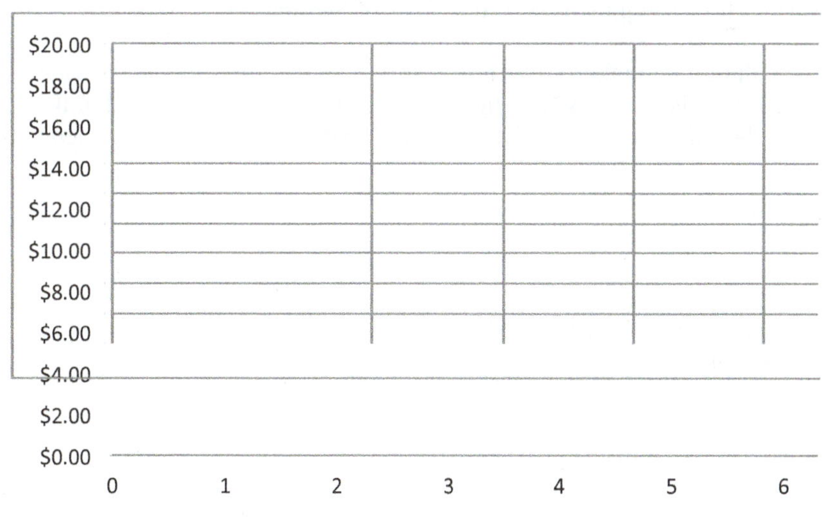

2. What do you notice about the pattern that the points on your scatter diagram are making? Use that observation about a pattern to connect the points, drawing a graph that represents the cost of a cab ride based on distance.

3. The point where any graph crosses the y axis is called the ***y* intercept** for the graph. Write the coordinates of the y intercept for the graph that you drew. More importantly, what information does the y intercept represent?

4. How much more will you pay for a 3 mile ride than for a 2 mile ride?

5. How much more will you pay for a 5 mile ride than a 2 mile ride? Explain why that makes sense based on your answer to Question 4.

6. Pick any two points on the graph that you drew, and subtract the second coordinates. Then divide the result by the difference of the first coordinates. What does the result represent about the cab ride?

In dividing the difference of the two costs by the difference of the two distances in Question 6, you found the rate of change of the cost as distance changes. When applied to the graph of a line, we call this number, which describes how steep the line is, the **slope** of the line.

7. The thing that makes a line a line is the fact that the slope never changes. We now know that the slope of a line describes the rate at which the y coordinate changes compared to the x coordinate. Based on this, how can you decide if the relationship between two quantities might be modeled well by a straight line?

8. Estimate the distance of a \$16 cab ride. You can use either the table or the graph, but make sure you explain how you got your answer.

9. If x represents the number of miles of a cab ride in Vegas, write an expression that represents the total cost of that ride.

Using Technology: Graphing an Equation With a Graphing Calculator

Starting with an equation with two variables x and y, with y alone on one side:

1. Press Y= to get to the equation editor screen.
2. Enter the side of the equation containing x, using X,T,θ,n for the variable.
3. Press WINDOW to get to the window screen.
4. Enter the lowest value you want to display along the x axis after Xmin, and the highest value you want to display along the x axis after Xmax. After Xscl, enter the distance that you want between each tick mark on the x axis. If you make this too small, the tick marks will be indistinct on your graph.
5. Enter the lowest value you want to display along the y axis after Ymin, and the highest value you want to display along the y axis after Ymax. After Yscl, enter the distance that you want between tick marks on the y axis.
6. Press GRAPH to display the graph. If you're not seeing a graph, you probably need to adjust the values you entered in the window screen.

See the Lesson 3-2 video in class resources for further information.

10. Use a graphing calculator to create both a table of values and a graph for the expression you wrote in Question 9. If you need a review of making a table, see the Using Technology feature in Lesson 2-8.

3-2 Group

Everyone knows that after you buy a car, in most cases its value decreases as it gets older. Suppose you bought a used car for $15,000 a few years ago, and its value has been decreasing at the rate of $2,000 per year since then. **Make sure you answer each question with a full sentence,** with punctuation and everything!

Time (yrs)	Value ($)
0	$15,000
1	$13,000
2	$11,000
3	$9,000
4	$7,000
5	$5,000

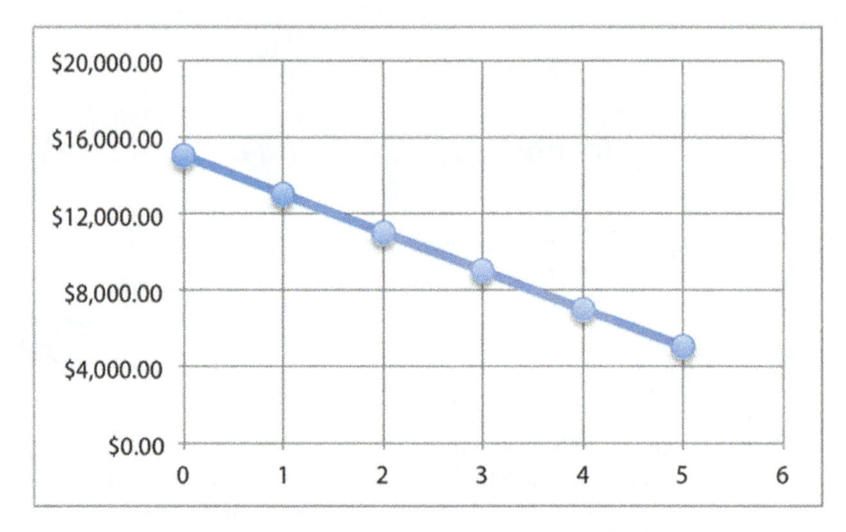

1. Find the y intercept of the graph and explain what it means. (Remember, the y intercept is a point, not a number.)

2. Find the slope of the line and explain what it means. In particular, what's the significance of the sign?

3. Estimate the value of the car after 42 months. (Careful about time units!)

4. Estimate the number of years it took for the car to reach a value of $6,000.

5. If x represents the number of years since the car was bought, write an expression that represents the value of the car.

6. Enter the expression from Question 5 next to **Y1** in the equation editor of a graphing calculator, then create a table and a graph.

> **Math Note**
>
> The process of an object losing value as time passes is called **depreciation.** This concept is used by businesses in calculating net assets for tax purposes.

In Lesson 2-5, we studied the profit made by a small business that sells snow cones. Let's examine a graph that describes the company's profit (y) in terms of the number of snow cones sold (x). Again, **answer each question with a full sentence** or face consequences too dire to mention in polite company. 😃

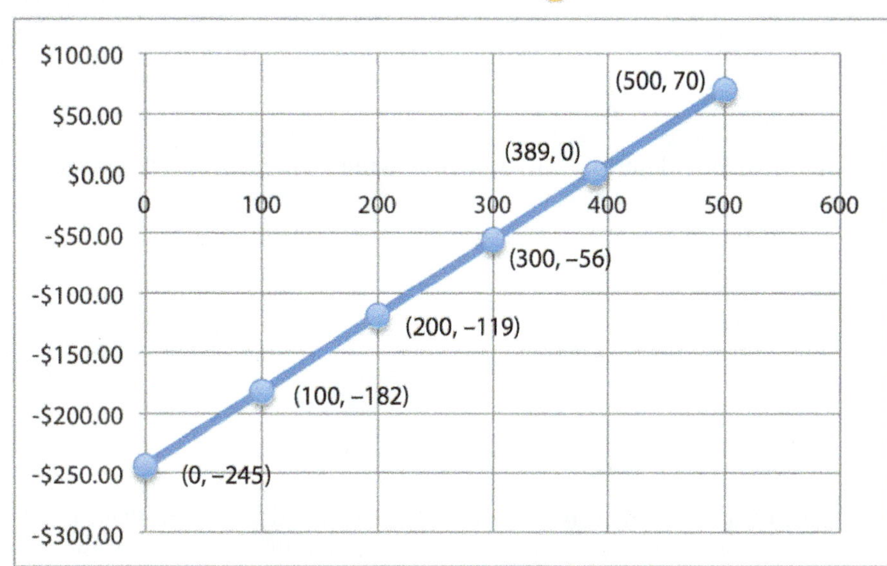

7. Find the y intercept of the graph and explain what it means.

8. A point where a graph crosses the x axis is called an ***x* intercept** of the graph. Find the x intercept of this graph and explain what it means.

9. Find the slope of the line and explain what it means.

10. Estimate the net profit if 150 snow cones are sold. What's the significance of the sign?

11. Estimate the number of snow cones that need to be sold to make a profit of $50.

12. Give a general description of how to find the slope of a line, and why it represents a rate of change. Be specific!

13. Give a general description of how to find the y intercept of a graph, and the x intercept as well. Include a definition in your own words of what the y and x intercept of a graph are.

14. Make up a practical problem that could correspond to each of these graphs. Include as much detail as possible, and make sure to pay close attention to the sign of the y coordinate. Be creative!

a.

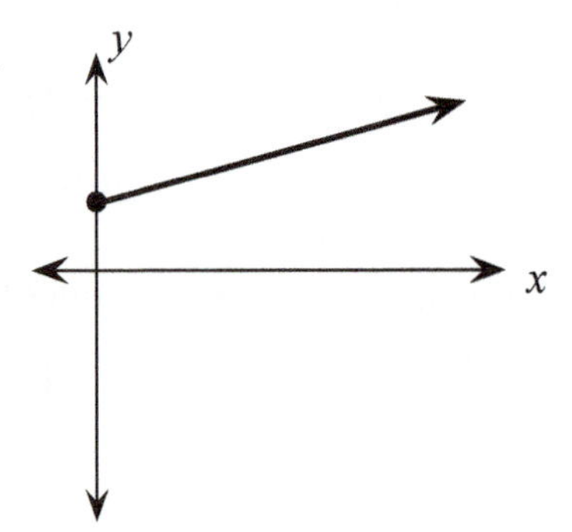

b.

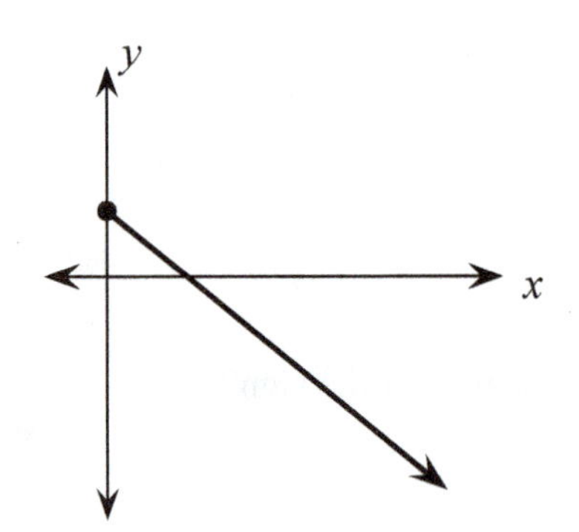

c.

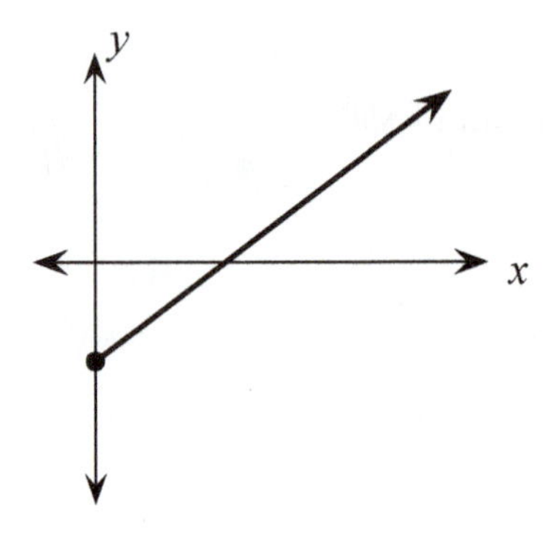

3-2 Portfolio

Name ______________________________

Check each box when you've completed the task. Remember that your instructor will want you to turn in the portfolio pages you create.

Technology

1. □ Complete the Applications portion of this lesson before tackling the Technology portion. Create a scatter diagram in Excel for the value of the industrial machine in the Applications section. Include values up to the point where the machine is worth nothing. Use the Straight Marked Scatter feature. A template to help you get started can be found in the online resources for this lesson.

Skills

1. □ Include any written work from the online skills assignment along with any notes or questions about this lesson's content.

Applications

1. □ Complete the applications problems.

Reflections

Type a short answer to each question.

1. □ Explain in your own words what the slope of a line is. You should discuss both what it means graphically, and what the practical significance is.
2. □ What are the intercepts of a graph?
3. □ Take another look at your answer to Question 0 at the beginning of this lesson. Would you change your answer now that you've completed the lesson? How would you summarize the topic of this lesson now?
4. □ What questions do you have about this lesson?

Looking Ahead

1. □ Read the opening paragraph in Lesson 3-3 carefully, then answer Question 0 in preparation for that lesson.

3-2 Applications

Name ______________________________

A manufacturing company buys a new stamping machine for $28,000. The maker of the machine informs the company's CEO that on average, it depreciates in value according to the schedule shown in the table.

Months	Value
0	$28,000
6	$24,500
12	$21,000
18	$17,500

1. If the depreciation continues at the same rate, how long will it take until the machine has no value?

2. Based on the pattern you see in the table, how do you know that the graph will be a straight line?

3. Pick an appropriate scale for each axis, and graph the value of the machine until it reaches zero. Use graph paper. Use the number of months as first coordinate.

4. Find the slope of the graph and explain what it means.

5. Find the intercepts of the graph, and describe what each means.

6. If we use the letter x to represent the variable number of months, write an expression that represents the value of the machine.

7. Use your expression from Question 6 to find when the machine has no value, and compare to the graph. Did you get the same result?

Lesson 3-3 Planning a Pizza Bash

Learning Objectives

- ☐ 1. Connect tables and graphs to solving equations.
- ☐ 2. Write equations of lines by recognizing slope and y intercept.

Leadership is solving problems.
– Colin Powell

In Lesson 2-9, we studied what it means to solve an equation, and learned a bit about the process of solving equations. Now it's time to turn our attention to something that I believe is important to each and every one of us: pizza. Or, more specifically, buying pizza under certain financial limitations, a position that most of us are pretty familiar with. We know that when a quantity has a constant rate of change, a graph representing that quantity will be a straight line. In this lesson, we'll practice writing equations that represent quantities with constant rates of change, and we'll see how the equation-solving techniques we learned can help us to solve problems involving those quantities.

0. After reading the opening paragraph, what do you think the main topic of this section will be?

3-3 Group

A student org you're involved in is planning a year-ending pizza party to celebrate another successful year on campus. As chair of the planning committee, you've wisely worked a deal with a local pizza place to provide large pizzas at $8 each. The budget will allow at most $150 to cover the food, and of course you'd like to know how many pizzas you can get and stay under budget. We'll attack the problem in several different ways.

1. Use a numerical calculation. Include all details, and don't forget to write the number of pizzas you can buy.

2. Complete the table, then use it to estimate the number of pizzas you can buy.

Pizzas Bought	Total Cost
0	$0
5	
10	
15	
20	
25	
30	

3. Use the information from the table to draw a graph with the number of pizzas on the x axis and the total cost on the y axis. Make sure you label the scale on each axis. Use your graph to estimate the number of pizzas that can be bought for $150. Try to ignore the answer you found in Question 1, and use only the graph.

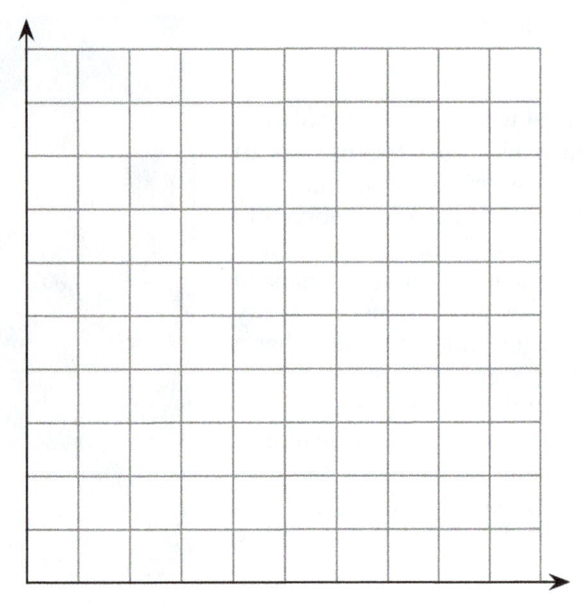

4. If we use p to represent the variable number of pizzas that can be ordered, what expression represents the total cost of buying p pizzas?

5. Use your answer to Question 4 to write an equation that represents the number of pizzas you can order for $150. Then solve that equation, and use the result to decide on the number of pizzas you can buy. **Write your answer in a full sentence.**

6. What's the slope of the line you drew in Question 3? What does it represent?

7. What's the y intercept of the line? What does it represent?

Nice job so far, but did we forget to mention the $10 delivery charge? Sorry. You'll have to rework the problem, keeping in mind that we want to know how many pizzas we can get for $150 at $8 each, plus a flat fee of $10 for delivery.

8. Use a numerical calculation. Include all details, and don't forget to write the number of pizzas you can buy.

9. Complete the table, then use it to estimate the number of pizzas you can buy.

Pizzas Bought	Total Cost
0	$10
5	
10	
15	
20	
25	
30	

10. Use the information from the table to draw a graph with the number of pizzas on the x axis and the total cost on the y axis. Don't forget to label a scale on each axis. Use your graph to estimate the solution to the problem.

11. Write and solve an equation that represents the number of pizzas you can buy with the delivery charge added. **Write the solution to the problem in a full sentence.**

12. What is the slope of the line you drew? What does it represent?

13. What is the y intercept of the line you drew? What does it represent?

3-3 Class

1. Suppose that you're in desperate need of some quick cash, and a local restaurant owner offers you twenty bucks to stand on a street corner and hand out coupons to her place. You also get a $0.50 bonus for every one of the coupons that gets turned in to the restaurant on that day.

a. How much will you get paid if nobody turns in the coupon?

b. If 20 coupons get turned in, how much will your bonus be?

c. If x coupons get turned in, how much will your bonus be?

2. Use your answers to Question 1 to write an equation that describes the total amount of money you'll make (y) if x coupons are turned in.

There's a very useful lesson to be learned from Questions 1 and 2.

- The $0.50 bonus you get for each coupon turned in is the rate of change of your total pay as the number of coupons turned in increases. And we know that a constant rate of change for some quantity represents the slope of a line when you graph that quantity.
- The $20 is the amount you'll get paid if none of the coupons get turned in. So zero coupons corresponds to $20, and the point (0, 20) would be on the graph describing total pay. In fact, it would be the y intercept.
- The equation describing total pay looks like $y = 0.5x + 20$, where x is the number of coupons turned in. Notice that the **coefficient** of x, which is the number x is multiplied by, is the slope of the line. Also, the **constant term,** which is the term without a variable, is the second coordinate of the y intercept.

Conclusion?

Writing the Equation of a Line

When a line has slope m and the second coordinate of its y intercept is the number b, then the equation of that line is $y = mx + b$.

3. A small business had just 12 employees when they opened their doors, but have been adding employees at the rate of 4 per month since then. Write a linear equation that describes the number of employees in terms of the number of months after the company was founded. How long will it take until they have 68 employees?

3-3 Portfolio

Name ______________________________

Check each box when you've completed the task. Remember that your instructor will want you to turn in the portfolio pages you create.

Technology

1. □ Use Excel to create a table and graph for both of the pizza party problems in the Group portion of this lesson. Make sure that you use a formula for calculating the costs. For each formula, explain how that formula relates to the equation that describes the cost of the pizzas. Type your explanation in the text box which is included in the template that you can find in the online resources for this lesson. A template to help you get started can be found in the online resources for this lesson.

Skills

1. □ Include any written work from the online skills assignment along with any notes or questions about this lesson's content.

Applications

1. □ Complete the applications problems.

Reflections

Type a short answer to each question.

1. □ Describe the connection between the graph of a line and how we use the line's equation to solve problems like the pizza party problem. How would you find the solution on a graph?
2. □ Explain how to find the equation of a line when you know the slope and the y intercept.
3. □ Take another look at your answer to Question 0 at the beginning of this lesson. Would you change your answer now that you've completed the lesson? How would you summarize the topic of this lesson now?
4. □ What questions do you have about this lesson?

Looking Ahead

1. □ Do an Internet search for "exchange rates" and find the value of one U.S. dollar in Russian rubles today. You'll need this number at the beginning of Lesson 3-4.
2. □ Read the opening paragraph in Lesson 3-4 carefully, then answer Question 0 in preparation for that lesson.

3-3 Applications Name ______________________________

A movie download service has a $14 monthly fee for membership; it then costs $3 to rent each movie.

1. Make a table that shows the relationship between the number of movies rented in a month, and the total fee. Vary the number of movies, but don't go past $80 spent.

Number of movies	Total cost

2. Draw a graph using the information from your table. Make sure to mark a scale on each axis, and include a verbal label that shows the quantity represented on each axis.

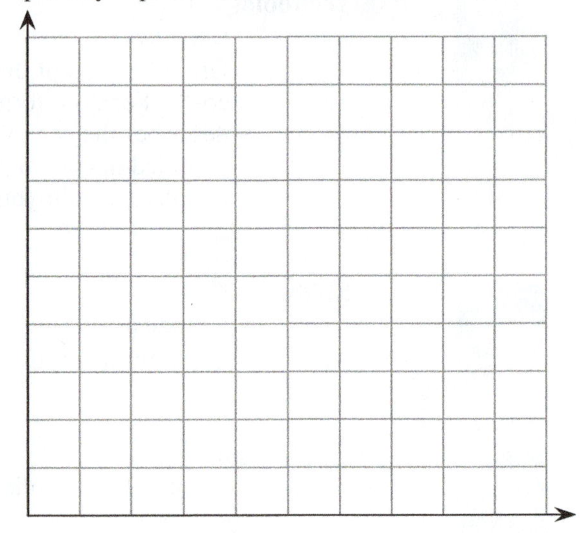

How many movies could you rent in a month if you don't want to spend more than $40? We'll tackle this problem in Questions 3-5.

3. Solve the problem using a numerical calculation. Show all details.

4. Write and solve an equation that solves the problem.

5. Write a full sentence that answers the question. More detail, as usual, is better.

6. What is the slope of the line in your graph? What does it represent?

7. What is the y intercept of the line? What does it represent?

Lesson 3-4 Big Mac Exchange Rates

Learning Objectives

☐ 1. Identify quantities that vary directly.

☐ 2. Write and use variation equations.

Decide what you want, decide what you are willing to exchange for it. Establish your priorities and go to work.
– H.L. Hunt

One of the most confusing things about traveling abroad is different currency. Since you're so familiar with dollars, you instantly have an idea of how cheap or expensive something is when the price is in dollars. What if I told you that a Big Mac costs about 75 rubles in Russia and 43 kroner in Norway? Those certainly sound like awful prices, but unless we know what a ruble or a kroner is worth, we have no idea how much or little that actually is. In this lesson, we'll use exchange rates between currencies to study the topic of variation, which can be used to model the connection between many useful quantities.

0. After reading the opening paragraph, what do you think the main topic of this section will be?

3-4 Class

An **exchange rate** is a number that describes how much of one currency you can trade for another currency. For example, if the U.S. exchange rate for Canadian currency is 1.2, it means that you could trade one U.S. dollar for $1.20 Canadian. When travelers talk about how expensive or cheap a certain country is, it's often a reflection of the exchange rate. The Big Mac costs mentioned earlier? The average cost in the U.S. in July 2012 was $4.33. In Russia it was just $2.29, and in Norway the cost was over $7.

1. Using the exchange rate you found online in the Looking Ahead portion of Lesson 3-3, complete the table that shows the relationship between the number of U.S. dollars and Russian rubles that can be exchanged for those dollars.

U.S. Dollars	Russian Rubles
0	
10	
20	
30	
40	
50	

2. Draw a graph based on your table. Make sure to mark a scale on each axis, and include a verbal label that shows the quantity represented on each axis.

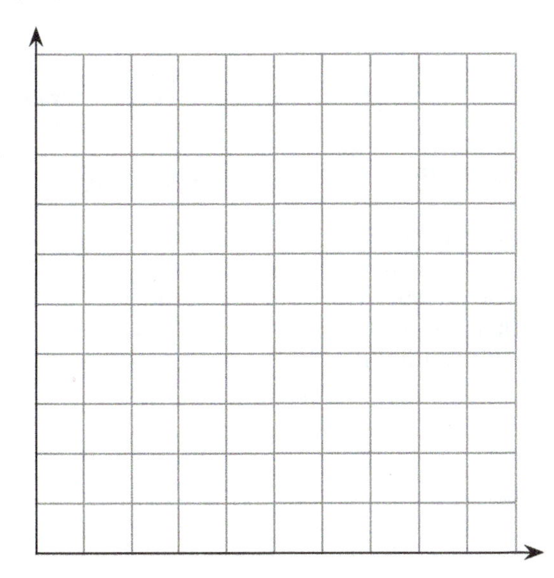

3. Find the slope of the line you graphed. What does it represent?

4. Write an equation using x and y that calculates the number of rubles (y) that you can trade x American dollars for.

5. Use your equation to find the current cost of a Big Mac in Russia, using \$2.29 as the equivalent in American dollars.

6. Use your graphing calculator to verify the table and graph from Questions 1 and 2.

Three things should be very clear about exchange rates:

- If the amount of dollars goes up, so does the amount of rubles.
- If the amount of dollars goes down, so does the amount of rubles.
- Zero dollars will get you zero rubles.

These three conditions are usually part of **direct variation**. Algebraically, the thing that shows us that two quantities x and y vary directly is an equation of the form $y = kx$, where k is some real number. (So if your equation from Question 4 doesn't look like that, you might think about changing it.)

Direct Variation

The following are various ways of illustrating what it means to say that two variable quantities vary directly.

Verbally

The quantity y varies directly as the quantity x, and the constant of variation is k.

Algebraically

$y = kx$

Example: $y = 3x$

Numerically

x	$y = 3x$
1	3
2	6
3	9
4	12
5	15

Graphically

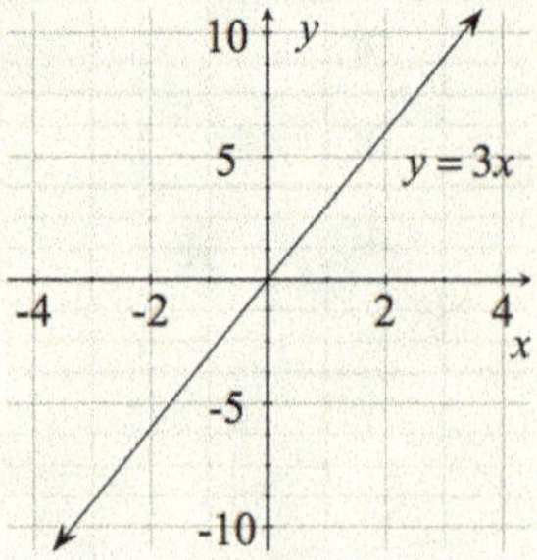

$y = 3x$

3-4 Group

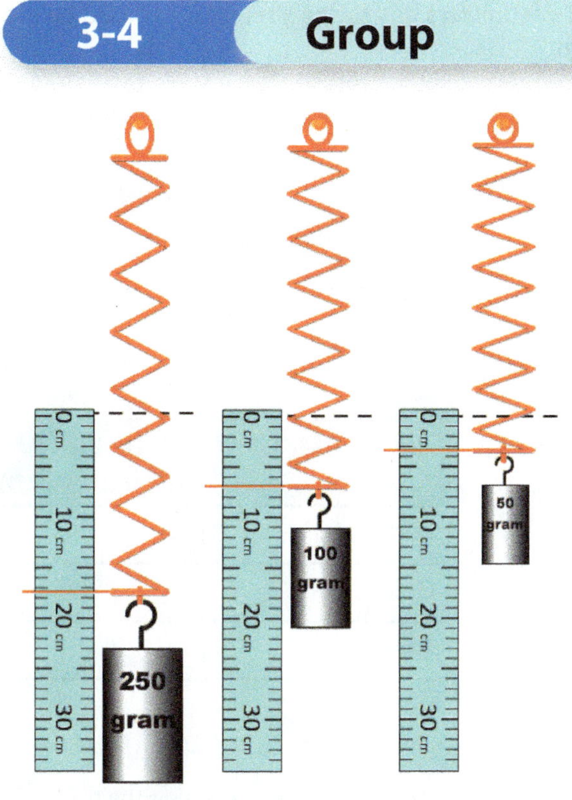

http://phet.colorado.edu/sims/mass-spring-lab/mass-spring-lab_en.html

Obvious fact: When you hang weights from a spring, the spring will stretch further for heavier masses. A cool simulation at the web page referenced below the diagram (linked in the online references for this lesson) allows you to measure the amount of stretch for various springs with masses of 50, 100, and 250 grams. The illustration displays the results for one spring. The dotted lines show the natural length of the spring, with no weight hanging on it at all.

1. Make a table that shows the relationship between the number of grams hung from the spring and the distance that it stretches. (Note that this isn't the stretched length: it's how far it stretched from its natural length.)

Mass (g)	Distance stretched (cm)
0	
50	
100	
250	

2. Draw the graph of a line that corresponds to the data in your table. Don't forget to label a scale on each axis.

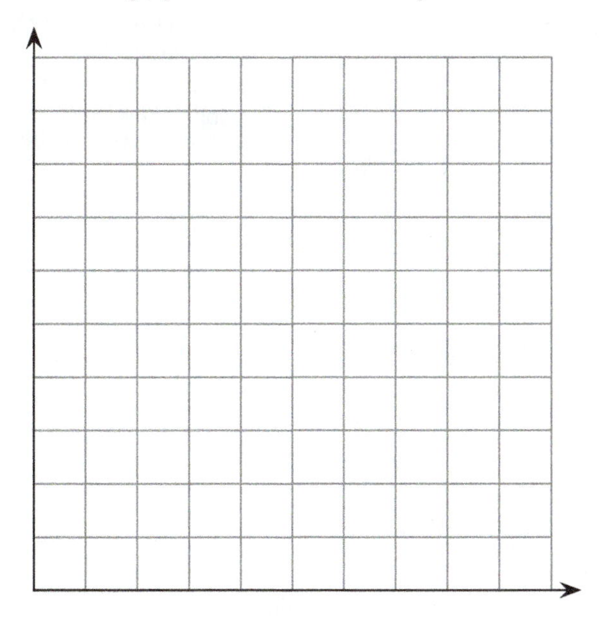

3. Find the slope of the line you graphed.

4. Write an equation using x and y that calculates the number of centimeters this spring stretches (y) based on the number of grams of mass (x) hung from the end of the spring.

5. Use your equation to calculate the amount of stretch for the 100 gram weight. Does it match your measurement?

6. Use your equation to find the mass of each of the colored weights.

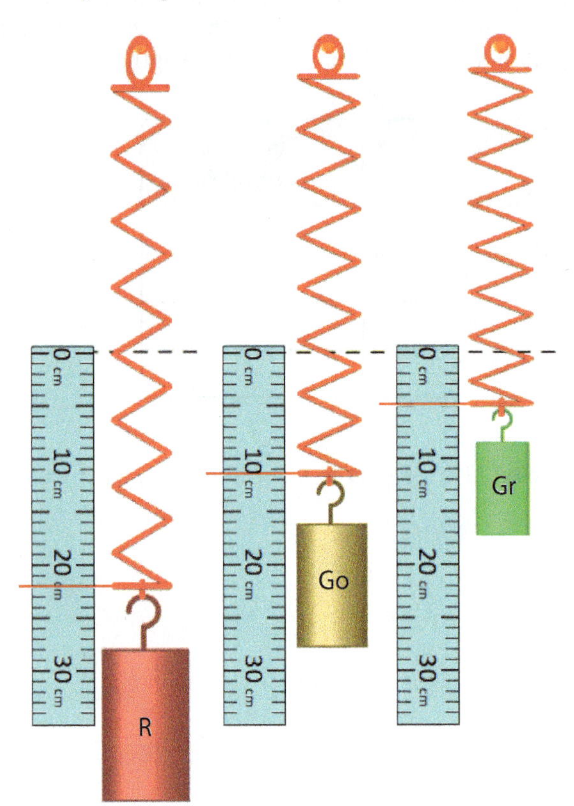

Color	Mass (g)	Distance stretched (cm)
Red		
Gold		
Green		

Math Note

The fact that the distance stretched by a spring varies directly as the weight hung on it is known in physics as Hooke's Law. If you've ever jumped on a trampoline, you've seen Hooke's Law in action: it's used in trampoline design.

7. What is the relationship between the slope of the line and the variation constant k in your equation? Will this always be the case for direct variation problems? Why?

8. When you know that two quantities vary directly, how many corresponding values do you actually need to find the value of k in the variation equation $y = kx$?

9. Once they reach cruising altitude, for efficiency, most planes fly at a constant speed on long flights. If one plane covers 1,275 miles in 3 hours, what is the rate the plane flies at in miles per hour?

10. Write an equation that describes the distance covered by the plane (y) after a certain number of hours (x). Think about what the rate you find in Question 9 really means.

11. Based on the form of your equation, what can you say about the relationship between distance traveled by a plane and the time it's been flying when the speed is constant? (If you're not sure, look back at the colored box at the end of the Class portion of this lesson.)

12. Use your equation and a graphing calculator or Excel to complete the table of values shown below.

Hours, x	Distance, Y_1
0	
1.5	
3	
4.5	
6	
7.5	
9	

13. Divide each distance in your table by the corresponding number of hours. What do you notice?

Bonus Question: Based on your answer to Question 13, how does direct variation connect to the concept of proportions that we studied in Lesson 2-9?

14. Earlier, we observed that as the number of rubles goes up, so does the number of dollars. And we also saw that as the mass of a hanging object goes up, so does the amount the spring stretches. Is that always the case? If the constant of variation is negative, what is the effect on y if x increases? Explain. Studying the generic equation that describes direct variation will help.

In Questions 15-17, decide if the graph illustrates a situation where y varies directly as x, and explain your reasoning.

15.

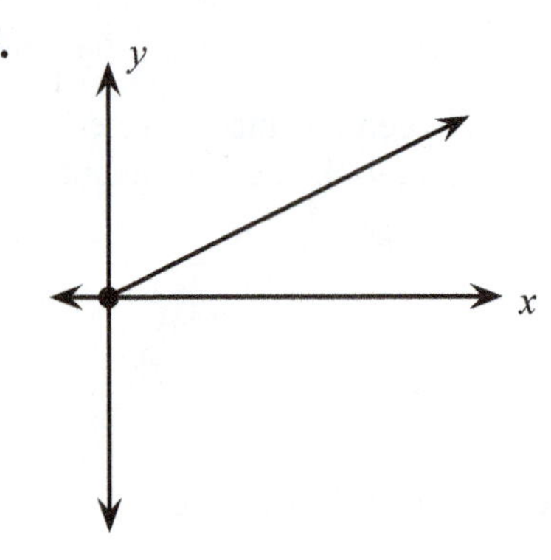

16.

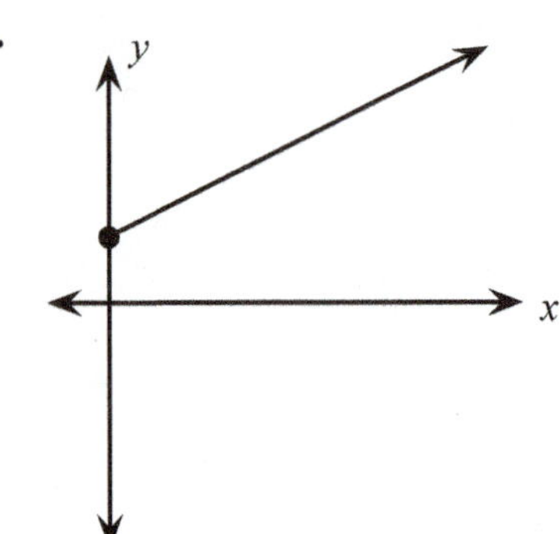

17.

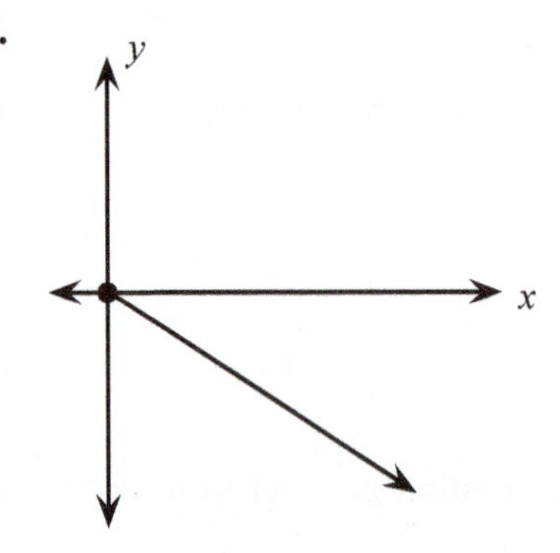

3-4 Portfolio

Name ______________________________

Check each box when you've completed the task. Remember that your instructor will want you to turn in the portfolio pages you create.

Technology

1. □ As you know, when you copy and paste a formula in Excel, it changes the cell reference accordingly: If the original formula is =A1/B1 in cell C1, copying and pasting in C2 changes the formula to =A2/B2. But sometimes we want to have all copied calculations use the same cell in a formula. In that case putting a dollar sign in front of the row and column will "lock" the cell reference and keep it from changing when we paste. This is useful in building an Excel table to convert currency.

Look up the exchange rate for converting U.S. dollars to the currency for a country you would like to visit. Type that rate in cell E1. Then create a table and graph similar to the ones in the Class portion of this lesson. Make sure you include the name of the country and the website where you got your exchange rate. A template to help you get started can be found in the online resources for this lesson.

	A	B	C	D	E
1	US Dollar	Converted Currency		Exchange Rate:	
2	0	=A2*E1		Country Name:	
3	10			Website used:	
4	20				
5	30				
6	40				
7	50				

Skills

1. □ Include any written work from the online skills assignment along with any notes or questions about this lesson's content.

Applications

1. □ Complete the applications problems.

Reflections

Type a short answer to each question.

1. □ How would you decide whether or not two quantities vary directly?

2. □ Take another look at your answer to Question 0 at the beginning of this lesson. Would you change your answer now that you've completed the lesson? How would you summarize the topic of this lesson now?

3. □ What questions do you have about this lesson?

Looking Ahead

1. □ Read the opening paragraph in Lesson 3-5 carefully, then answer Question 0 in preparation for that lesson.

3-4 Applications Name ____________________

1. Use the Internet to find and write the current exchange rate for converting the U.S. dollar to the Chinese yuan.

2. Complete the table using the current exchange rate.

U.S. Dollar	Chinese Yuan
0	0
20	
40	
60	
80	
100	

3. Write an equation that will convert U.S. dollars to Chinese yuan.

4. Which is the independent variable in your equation? Which is the dependent variable? Describe each using the terms input and output.

5. Use the Internet to find and write the current exchange rate for converting the Chinese yuan to the U.S. dollar.

6. Complete the table using the current exchange rate.

Chinese Yuan	U.S. Dollar
0	0
20	
40	
60	
80	
100	

3-4 Applications Name ______________________________

7. Write an equation that will convert Chinese yuan into U.S. dollars.

8. Which is the independent variable in your equation? Which is the dependent variable? Describe each using the terms input and output.

9. The average cost of a Volkswagen in China is about 150,000 yuan. How much is that in dollars? Use your equation from Question 7 and show all work.

10. It costs about $1,400 to fly from New York to Beijing. How much is that flight in yuan? Use your equation from Question 3 and show all work.

11. The Chinese government has been accused of manipulating the value of its currency, keeping it artificially low in order to give the country an export advantage. Some economists feel that the yuan is valued at 20% less than it should be compared to the U.S. dollar. If that's the case, what should the exchange rate be? How much would that affect the price of a Volkswagen in China (see Questions 5 and 9)?

Lesson 3-5 The Effects of Alcohol

Learning Objectives

- ☐ 1. Write an equation of a line that models data from a description, table, or graph.
- ☐ 2. Graph lines from a given equation.

As long as algebra is taught in school, there will be prayer in school.
– Cokie Roberts

It's no secret that alcohol consumption by college students is common, and that overindulgence can have many negative effects, in both the short and long term. When someone makes the choice to drink, they're affected as long as the alcohol remains in their bloodstream. There are many theories on how to hasten the sobering up process: black coffee, ice cold showers, greasy food, drinking lots of water ... . The truth is that none of these have any effect whatsoever on blood alcohol concentration (BAC). You can try every mythical treatment known to man, but at the end of the day the only thing that will bring down BAC is time. So far, we've studied a wide variety of quantities that can be modeled using linear equations. In this lesson, we'll use the time required to sober up, and other quantities, to study writing linear equations that model situations in greater depth.

0. After reading the opening paragraph, what do you think the main topic of this section will be?

3-5 Class

Blood alcohol concentration is a percentage of alcohol in the blood stream. The legal limit for operating a motor vehicle in most states is 0.08: this means that 0.08% of a person's blood is alcohol. The table below is borrowed from WebMD. It shows the ways that alcohol can affect a person at different BAC levels.

The Effects of Drinking Alcohol

Estimated blood alcohol concentration (BAC) %	Observable effects
0.02	Relaxation, slight body warmth
0.05	Sedation, slowed reaction time
0.10	Slurred speech, poor coordination, slowed thinking
0.20	Trouble walking, double vision, nausea, vomiting
0.30	May pass out, tremors, memory loss, cool body temperature
0.40	Trouble breathing, coma, possible death
0.50 and greater	Death

1. After being arrested for driving under the influence, a well-known starlet was found to have three and a half times the legal limit of alcohol in her blood. What was her blood alcohol concentration?

2. Explain exactly what your answer to Question 1 means.

3. Describe some of the effects the arresting officers may have been able to observe at the time of the arrest.

4. Suppose that a person of legal drinking age starts drinking at 9 PM and continues downing 2 drinks per hour steadily until 1:30 AM. If he's of average size, his BAC would peak at around 0.16 at 2 AM. Studies have shown that the typical person's BAC decreases by 0.015 per hour after he or she stops drinking and blood alcohol peaks. Write a linear equation that describes the drinker's BAC in this case, with y representing BAC and x representing the number of hours after 2 AM. (Hint: How many hours after 2 AM is 2 AM?)

5. When using an equation to model a situation, it's important to recall what specific information the equation provides. Restate your answer to Question 4 in the form of a statement like this: "If x represents ___________, then y = _______________ represents _____________."

6. We've learned about several features of graphing calculators that allow us study the information provided by an equation: substituting in values for variables, making a table, and making a graph come to mind. Using these features, discuss whether or not the equation you wrote tells the story that it's supposed to. More detail is always better!

7. Let's ignore the equation we're modeling BAC with for just a moment. Based on the description of the situation in Question 4, what should the value of y be when x is 0? What about when x is 1? Explain how you got your answers.

8. Do the outputs (y values) corresponding to inputs $x = 0$ and $x = 1$ match your answers from Question 7?

9. What would our friend's blood alcohol concentration be at 4 AM? (Remember, eating greasy food won't change it!)

10. How many hours would it take for his BAC to drop to the legal driving limit of 0.08? (He can take a cold shower if he likes, but it won't help.) Set up and solve an equation to answer this question, please.

11. What information would you need in order to draw the graph of your equation? Be specific.

12. Graph the line on the coordinate system provided. Make sure that you choose and label an appropriate scale on each axis.

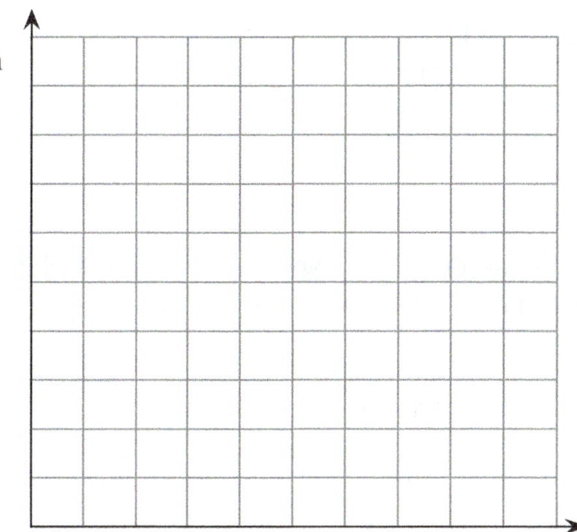

13. What is the slope of the line? What does it mean?

14. What is the y intercept of the line? What does it mean?

15. What is the x intercept? What does it mean?

3-5 Group

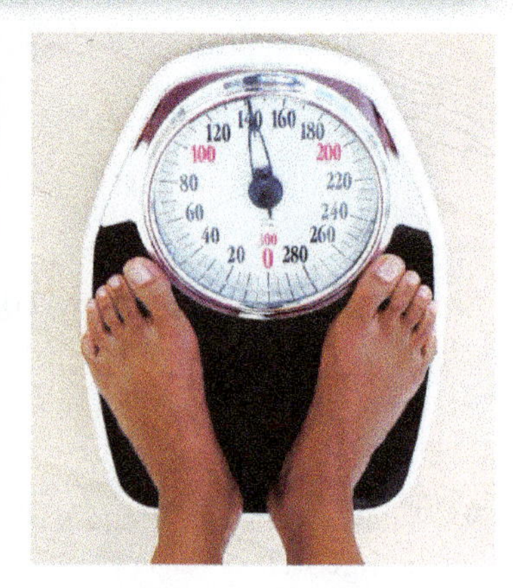

If you're interested in weight loss, there certainly isn't a shortage of information, diet plans, supplements, and flat-out gimmicks available to you. But the simple truth is this: gaining or losing weight ultimately comes down to one thing. If you consume more calories than you burn, you'll gain weight, and if you consume less calories than you burn, you'll lose weight. In this activity we'll study some aspects of weight change.

1. Maria weighed 130 lbs when she began college. She then started gaining weight at the rate of 0.25 lbs per month. Write a linear equation representing Maria's weight (y) in terms of months after she started college (x).

2. Restate your answer to Question 1 in the form of a statement like this: "If x represents ____________, then $y =$ ______________ represents ____________."

3. If Maria continues this modest-sounding weight gain, how much will she weigh after three years of college?

4. How long will it take our buddy Maria to reach 160 lbs at this rate?

5. What information would you need in order to draw the graph of your equation? Be specific.

6. Draw the graph of your equation, including (as usual) an appropriately chosen scale on each axis.

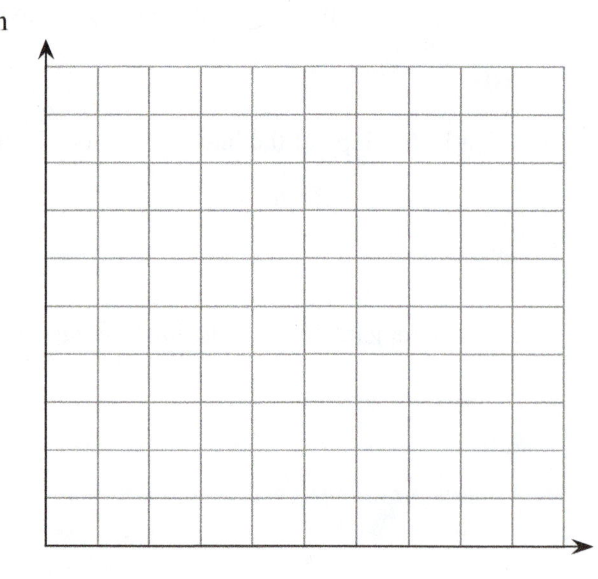

7. Find the slope of the line and explain what it means.

8. Find the y intercept and explain what it means.

Now let's study what it takes to lose weight. In Lesson 2-7, we saw that an average person can lose one pound if she burns 3,500 calories more than she consumes. Let's say that after reaching the 160 pound plateau, Maria has had enough and decides to work on losing some weight. The table shows Maria's weight if a certain number of calories are burned.

Calories burned (thousands)	Weight
0	160
14	156
28	152
42	148
56	144

9. Find the slope of a line based on the data in the table. Use weight for y coordinates and thousands of calories burned as x coordinates.

10. Write a linear equation that describes Maria's weight (y) in terms of thousands of calories burned (x). Your answer to Question 9 will help.

11. Restate your answer to Question 10 in the form of a statement like this: "If x represents ____________, then $y =$ ______________ represents ____________."

12. Use your equation to find Maria's weight if she burns 50,000 calories above what she consumes.

13. Use your equation to find how many extra calories Maria would have to burn to get back to the weight at which she started college.

14. According to the Mayo clinic website, a woman of Maria's size would burn about 606 calories when running for an hour. How many hours of running will Maria need to put in to get back to her original weight?

15. Explain what the slope you found in Question 9 means in this situation..

16. Find the y intercept of the line and explain what it means.

17. Find the x intercept of the line and explain why it's totally irrelevant to the situation that we're modeling.

18. Based on this entire group activity, what would have been Maria's best approach when starting college?

Math Note

Interested in overcoming the famous "freshman fifteen"? A Google search for that phrase yields some very informative and useful results.

3-5 Portfolio

Name ______________________________

Check each box when you've completed the task. Remember that your instructor will want you to turn in the portfolio pages you create.

Technology

1. ☐ In this Portfolio, you'll need to complete the Applications portion before Technology. And you'll have to be creative! Create a table and graph from the tuition bill problem. Try to use as many features that we've learned in Excel as you can. Formulas, scroll bars and $ signs to lock cell locations would impress me and, more importantly, your instructor.

Skills

1. ☐ Include any written work from the online skills assignment along with any notes or questions about this lesson's content.

Applications

1. ☐ Complete the applications problems.

Reflections

Type a short answer to each question.

1. ☐ Describe what you learned about the time required to sober up from the Class portion of this lesson.
2. ☐ Describe what you learned about weight gain and loss from the Group portion of this section.
3. ☐ When a situation can be modeled by a linear equation, what information do you need in order to find an equation?
4. ☐ Take another look at your answer to Question 0 at the beginning of this lesson. Would you change your answer now that you've completed the lesson? How would you summarize the topic of this lesson now?
5. ☐ What questions do you have about this lesson?

Looking Ahead

1. ☐ Read the opening paragraph in Lesson 3-6 carefully, then answer Question 0 in preparation for that lesson.

3-5 Applications Name ______________________________

1. The table below shows the cost of tuition at a local community college based on the number of credit hours a student is registered for. Plot points corresponding to the given information, with credit hours on the x axis and cost on the y axis.

Credit hours	Cost
12	$1,435
15	$1,750
16	$1,855

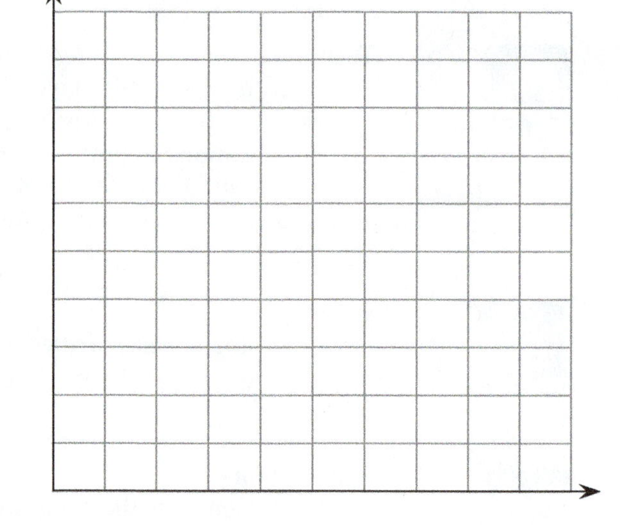

2. Explain why a linear equation seems like a reasonable choice for modeling this data.

3. Graph a line through the points you plotted, including all the usual important labeling.

4. Find the slope of the line and explain what it means.

5. Find the y intercept of the line and explain what it means.

6. Use the form $y = mx + b$ to write an equation for the line. You can either use your answer to Question 5, or use one of the points from the table to set up and solve an equation for b.

7. Write your equation in the form of a statement: "If x represents _____________, then y = _______________ represents _____________."

8. Use your equation to find the cost of taking 8 credit hours, and the number of credits taken by a student with a tuition bill of $2,170.

Lesson 3-6 The Great Tech Battle

Learning Objectives

- ☐ 1. Decide if two data sets are linearly related.
- ☐ 2. Find the line of best fit for data using spreadsheets and calculators.

One machine can do the work of fifty ordinary men. No machine can do the work of one extraordinary man.
– Elbert Hubbard

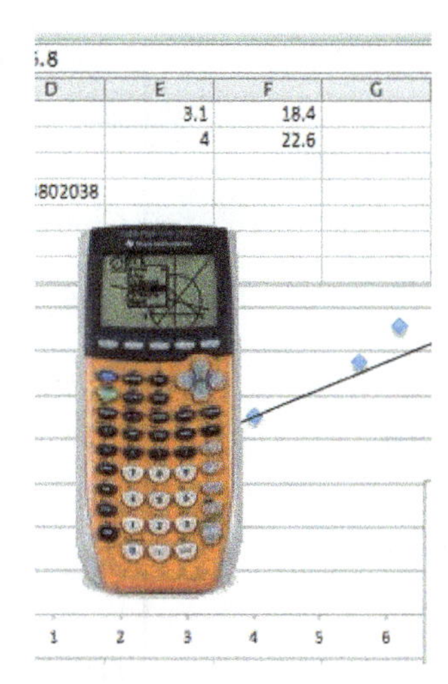

That quote says an awful lot about the dangers of too much reliance on technology, a peril with the potential to affect every modern student. The most interesting thing about it, though, is that it was written in 1903! When used correctly in math, technology allows us to focus more on understanding and interpretation, and less on computation. That's great. But when you try to use any technology as a substitute for thinking, nothing good is likely to happen. In this lesson, we'll learn how technology can help us to decide when a given data set is likely to be modeled well with a linear equation. Better still, calculators and spreadsheets will be able to find the best linear model for such situations, freeing us to focus on interpreting the model and using it to further study the data.

0. After reading the opening paragraph, what do you think the main topic of this section will be?

3-6 Class

In Lesson 3-5, we examined the relationship between calories burned and weight loss. While we needed to study data to model the exact nature of the relationship, it doesn't exactly take a genius to figure out that the two quantities are related in some way. There are many quantities that *seem* like they might be related, but it's not clear that they really are, or if so, how. So it would be nice to have a way to study relationships (if they even exist) between data sets. Fortunately, the lovely folks that program spreadsheets and graphing calculators have done most of the heavy lifting for us. Still, in our continuing quest to not rely on technology more than our brains, it's a good idea to think about when quantities are likely to be related in some way.

1. Decide if you think the two quantities are likely to be related in some way, and explain your answer.

a. Dollars spent by a political candidate and the number of votes he gets

b. An adult test subject's height and their IQ

c. A person's height and their shoe size

d. The age of a car and its resale value

2. Is there a relationship between the total fat in a fast food sandwich and the total calories? Begin your study of this question by drawing a scatter plot based on the table.

Sandwich	Grams of fat	Calories
Hamburger	9	260
Cheeseburger	13	320
Quarter Pounder	21	420
Quarter Pounder with Cheese	30	530
Big Mac	31	560
Arch Special	31	550
Arch Special with Bacon	34	590
Crispy Chicken	25	500
Fish Filet	28	560
Grilled Chicken	20	440
Grilled Chicken Light	5	300

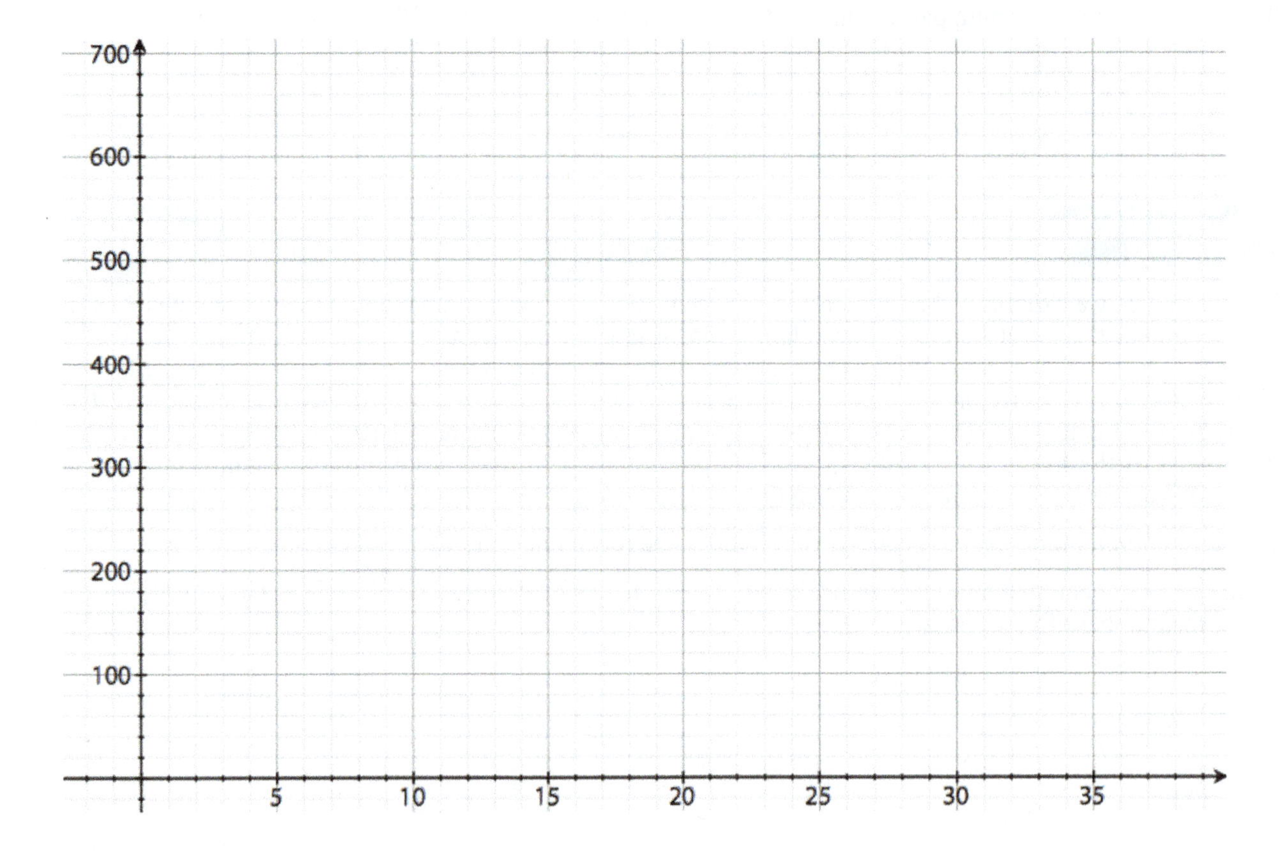

3. The points on your scatter plot don't form a straight line, but there should be a linear trend you can observe. Use a straightedge to draw a line that you think looks like the best fit for your scatter plot. Note that it's possible that the line that fits the data best doesn't go through any of the points.

Now that we've been able to draw a line that seems like a reasonable fit for data, it would be nice to find the equation of that line so we can do some calculations with it. We'll leave the sandwich problem just for a bit to review some algebra.

We've been using the form $y = mx + b$ to find the equations of lines so far. This worked well because b represents the second coordinate of the y intercept, and in every case we've either known or been able to find the y intercept. But what if we don't know the y intercept? We can develop a procedure that allows us to find the equation when we know ANY two points on the line.

A Procedure For Finding The Equation of a Line Through Two Points

1. Find the slope using our slope formula.
2. Substitute the slope, and the x and y coordinates of one of the points in for x and y in the equation $y = mx + b$. This results in an equation you can solve to find b.

4. Find the slope of the line connecting the points (5, 10) and (9, 30).

5. At this point, we know that the equation looks like $y = mx + b$, where m is the slope you found in Question 4. Write this equation with your slope, keeping the symbol b. (Remember, b isn't just any old second coordinate: it's the second coordinate of the point with first coordinate zero!)

6. We know that (5, 10) is on the graph of our line, so if we substitute 5 for x and 10 for y in the equation you wrote in Question 5, the resulting equation must be a true statement. Do that now, and you'll find that you have an equation you can solve to find b. Solve it!

7. Use your answer from Questions 5 and 6 to write the equation of the line connecting the points (5, 10) and (9, 30) in the form $y = mx + b$.

We now return to the scatter plot and line of best fit for the calorie/fat connection in fast food sandwiches.

8. Identify two points on the line of best fit that you drew on your scatter plot, and use them to find an equation for the line.

Math Note

To see if a point is a solution to an equation with two variables, substitute the first coordinate for x and the second coordinate for y. If the resulting equation is true, the point is on the line.

Using Technology: Finding the Line of Best Fit With a Graphing Calculator

To find a line of best fit for two data sets that appear to have an approximate linear relationship:

1. Press STAT ENTER to get to the list editor.
2. Enter the data set you want to use as inputs (x values) under **L1.**
3. Use the right arrow key ▸ to access **L2,** then enter the data set you want to use as outputs (y values). When entering data from a table, make sure you enter the second list in the same order as the first.
4. Turn on **Plot1** by pressing 2nd Y= 1. Set up the screen as shown below.
5. Press Y= and if the **Y** = screen isn't blank, move the cursor over any entered equations and press CLEAR.
6. Press ZOOM 9 which is the **Zoom Stat** option; this automatically sets a graphing window that displays all of the plotted points.
7. Press STAT followed by the right arrow key to access the **STAT CALC** menu, and choose the **LinReg** (for linear regression) option, which is choice 4. Then press ENTER to calculate the line of best fit.
8. Press Y= and enter the equation of the line of best fit, then press GRAPH to display the scatter plot along with the line of best fit.

See the Lesson 3-6-1 video in class resources for further instruction.

Plot1 Plot2 Plot3
On Off
Type:
Xlist:L1
Ylist:L2
Mark:

9. Use the procedure in the Using Technology box to find the equation of the line of best fit using a graphing calculator. Round to one decimal place.

10. Use your answer to Question 9 to predict the number of calories in a Whopper, which has 39 grams of fat. How does the answer compare to the number of calories predicted by your own line of best fit in Question 8?

When finding the line of best fit using a graphing calculator, the last line in the display should be a value for a quantity labeled r, which is known as the **correlation coefficient.** This number measures how well the line seems to fit the data. The closer r is to 1 (positive slope) or −1 (negative slope), the more accurately the data you entered can be modeled using a linear equation. Here are some examples, based on data from two of my classes last semester:

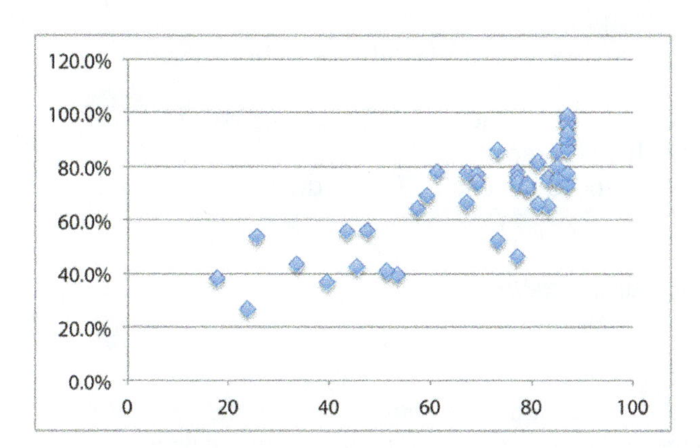

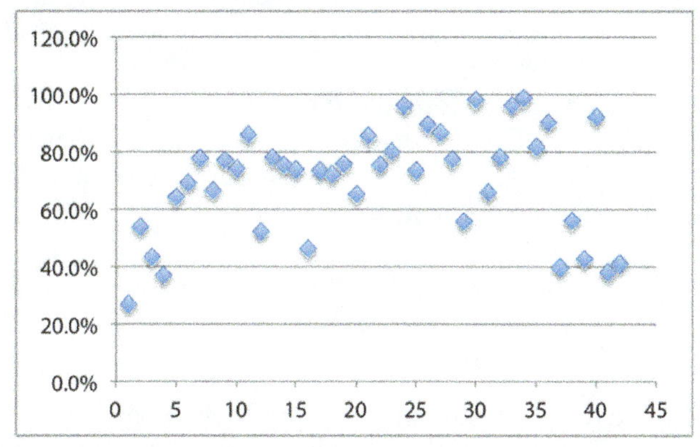

11. One of these scatter plots has a correlation coefficient of 0.19; the other has a correlation coefficient of 0.82. Discuss which you think is which, and why.

12. One of the scatter plots has each student's homework score for the semester on the x axis and the corresponding overall percentage in the course (from 0 to 100) on the y axis. The other again has overall percentage on the y axis, but along the x axis is the student's position alphabetically in the class. (The student whose name comes first alphabetically is 1, the next alphabetically is 2, and so on.) Which plot is which? Again, discuss your reasons.

13. What is the value of the correlation coefficient r for the fast food sandwich plot? Round to two decimal places. (If your calculator is not displaying r, press 2nd 0, then use the down arrow to scroll down and choose "DiagnosticOn" and try the calculation again.)

Using Technology: Finding the Line of Best Fit With a Spreadsheet

To find a line of best fit and correlation coefficient on a spreadsheet:

1. Enter the data in two columns and create a scatter plot.
2. Click on one of the points on the scatter plot, then choose "Add Trendline" from the **Chart** menu.
3. In the formatting dialog box that appears, click "Options", then click the checkboxes for "Display equation on chart" and "Display r^2 value on chart."
4. Calculate the square root of the r^2 value that appears on the chart. This is the correlation coefficient.

See the Lesson 3-6-2 video in class resources for further instruction.

14. Use the procedure in the Using Technology box to find the equation of the line of best fit and correlation coefficient using Excel. How does each compare to the results you got from your graphing calculator?

15. Based on all of the work done in this activity, how strong do you think the relationship is between the number of grams of fat in a fast food sandwich and the number of calories in it? Make sure you justify your answer.

An Important Note on Lines of Best Fit

In Question 10, we saw that the line of best fit for the number of grams of fat in a sandwich doesn't give the exact correct value for a Whopper. This will almost always be the case when finding a line of best fit for a data set with more than two values. There's a good reason we call it "the line of best fit," not "the line of perfect fit"! The equations we're finding are *models* for data that allows us to make approximations and predictions; they're not *exact representations* of data.

16. How close is the actual number of grams of fat in a Big Mac to the number predicted by the line of best fit? Is the prediction too high or too low?

3-6 Group

1. Do you think there's a strong connection between a person's shoe size and their height? Answer yes or no, explain why you feel that way, and make a guess as to the value of the correlation coefficient for these two data sets.

2. Fill out the first two columns in the chart for all of the students in your class, your instructor, and anyone else who happens to be walking by.

Shoe Size	Height (inches)	Predicted Height

3. Draw a scatter plot on graph paper with shoe size on the x axis and height in inches on the y axis.

4. Use a calculator or spreadsheet to draw a scatter plot for the data, then find and write the equation of the line of best fit.

5. Add the line to your hand graph.

6. Find the correlation coefficient for the data. How closely are the two quantities related?

7. Use your equation to predict the height of a person with size 17 shoes.

8. Use your equation and either the table feature on a graphing calculator or a spreadsheet to calculate the predicted heights for everyone in your chart based on shoe size.

9. What percentage of the people in your chart are within 2 inches of their predicted height? What can you conclude?

Olympic Men's Long Jump

The table to the right shows the gold-medal lengths for the men's Olympic long jump for selected years from 1900 to 2000. When dates are part of a data set, it's common to not use the year, but rather the number of years after the first date. This allows for smaller numbers in the calculation, which can reduce the amount of rounding error.

Year	x	Length (meters)
1900	0	7.18
1912	12	7.60
1920	20	7.15
1932	32	7.64
1948	48	7.82
1960	60	8.12
1972	72	8.24
1980	80	8.54
1996	96	8.50
2000	100	8.55

10. Use a calculator or spreadsheet to draw a scatter plot for the data with the number of years after 1900 on the x axis and length on the y axis. Then find the equation of the line of best fit. Round coefficients to three decimal places.

11. Find the slope of the line. What does it mean? More detail = Better.

12. The second table lists the length of the winning jumps for some years not in the table. Use your equation to find the predicted length of each winning jump.

Year	x	Length (meters)
1984	84	8.54
2004	104	8.59
2008	108	8.34
2012	112	8.31

13. How do the predictions compare to the actual jumps? What can you conclude?

14. Use your equation to predict the year that the men's long jump record will reach 9.0 meters. (Remember, the Olympics are only held every four years!)

15. Do an Internet search for "1968 men's long jump." After reading about the event that year, explain why we purposely left that year out of the table.

3-6 Portfolio

Name ______________________________

Check each box when you've completed the task. Remember that your instructor will want you to turn in the portfolio pages you create.

Technology

1. □ You kind of get a break on this one ... since this lesson is all about using technology, you'll take care of the technology part in doing the Applications portion. Cool!

Skills

1. □ Include any written work from the online skills assignment along with any notes or questions about this lesson's content.

Applications

1. □ Complete the applications problems.

Reflections

Type a short answer to each question.

1. □ What is a line of best fit for two sets of data?
2. □ Does the line of best fit provide useful information about every pair of data sets? Why or why not? Your answer should probably mention the correlation coefficient.
3. □ Take another look at your answer to Question 0 at the beginning of this lesson. Would you change your answer now that you've completed the lesson? How would you summarize the topic of this lesson now?
4. □ What questions do you have about this lesson?

Looking Ahead

1. □ Read the opening paragraph in Lesson 3-7 carefully, then answer Question 0 in preparation for that lesson.

3-6 Applications Name ______________________

The table to the right shows the average annual cost for tuition, room, and board at all colleges in the United States between the 2000 and 2010 school years.

School Year	x	Cost ($)	Increase
2000–2001	0	10,820	–
2001–2002	1	11,380	
2002–2003	2	12,014	
2003–2004	3	12,953	
2004–2005	4	13,793	
2005–2006	5	14,634	
2006–2007	6	15,483	
2007–2008	7	16,231	
2008–2009	8	17,092	
2009–2010	9	17,649	
2010–2011	10	18,497	

1. Calculate the increase over the previous year for each school year and put the value in the table. Based on the results, do you think we could model this data accurately with a linear equation? Why or why not?

2. Draw a scatter plot on the axes provided below. The x values are given in the table; use values from the Cost column as y values. Does the result seem to match your answer to Question 1?

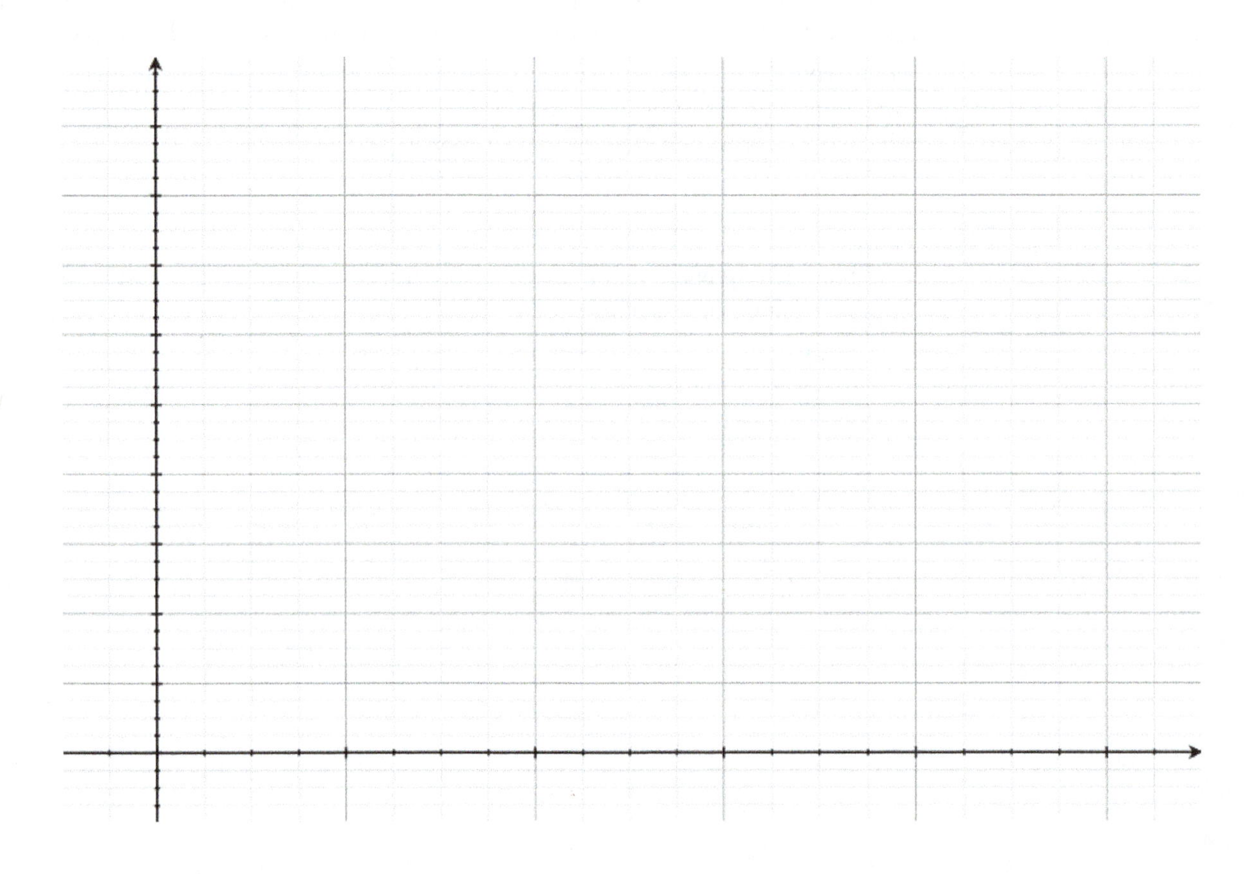

3-6 Applications Name ____________________

3. Use a graphing calculator or spreadsheet to create a scatter plot and find the line of best fit for the data, then add the line to your scatter plot.

4. What is the slope of the line? What does it mean?

5. Use your equation to predict total tuition, room, and board costs for the 2016–2017 school year.

6. Use your equation to predict total tuition, room, and board for the 1990-1991 school year. (Hint: What x value would correspond to ten years BEFORE 2000?)

7. Use your equation to predict the school year in which total tuition, room, and board will reach $25,000.

8. What is the correlation coefficient for the data? Are you surprised based on your answer to Question 1? Explain.

Lesson 3-7 If You Got a Problem, Yo I'll Solve It

Learning Objectives

- ☐ 1. Use equations to solve applied problems.
- ☐ 2. Use systems of equations to solve applied problems.
- ☐ 3. Illustrate problems with tables and graphs.

In the long run, we shape our lives, and we shape ourselves. The process never ends until we die. And the choices we make are ultimately our own responsibility.

– Eleanor Roosevelt

If a problem comes knocking at your door, you can hide behind the sofa and hope it goes away, or you can answer the door and see an opportunity. Many successful people will tell you that this mindset is what separates the winners from the pretenders in life. It's about taking personal responsibility for your own success, and seeing every obstacle as one more chance to achieve success. Throughout this unit, we've learned skills and strategies that can be used to write equations describing situations, and use those equations to solve problems. Now it's time to put these skills to the test, solving problems from a variety of settings. Think of them as opportunities to shine!

0. After reading the opening paragraph, what do you think the main topic of this section will be?

3-7 Class

1. Summarize Polya's problem solving strategy from Unit 2.

Solve each problem using both a numerical calculation and an algebraic equation.

2. How much would you save if you bought a laptop for $599 and got a 15% discount for ordering online?

Numerical Solution

Algebraic Solution

I'm using the letter ______ to represent the variable quantity ____________________________.

3. What percent would you save if you got a $50 mail-in rebate on a $299 cell phone?

Numerical Solution

Algebraic Solution

I'm using the letter ______ to represent the variable quantity ____________________________.

4. The FDA allows 13 insect heads for every 100 grams of fig paste used in fig cookies. (Legal disclaimer: I swear we're not making that up.) What's the maximum allowable number of insect heads you'd eat if you spent a week eating nothing other than fig cookies and downed 5,000 grams of fig paste?

Numerical Solution

Algebraic Solution

I'm using the letter ______ to represent the variable quantity __________________________.

5. While traveling on business, Eldrick bought a $50 prepaid international cell phone to call his new girlfriend Lindsey. Calls cost $0.07 per minute. How many minutes has he used if the display shows $32.15 remaining?

Numerical Solution

Algebraic Solution

I'm using the letter ______ to represent the variable quantity __________________________.

3-7 Group

1. With only a 200 point final remaining, Se Ri has earned 550 out of 600 points in her math course. A quick check of the syllabus reminds her that 80% is the minimum for getting a B.

a. What percent of the points has Se Ri earned so far?

b. If she earns a 50% on the final exam, what will her overall percentage be?

c. How many points does Se Ri need on the final to get a B in the course? Set up and solve an algebraic equation, and write your answer in the form of a sentence.

I'm using the letter ______ to represent the variable quantity ____________________________.

2. A new TV show premieres amid a great deal of promotion and hype, and 17.3 million viewers tune in for the first episode, which kinda stinks. For episode 2, only 8.5 million people watch, and after that the number of viewers starts to decline at a constant rate. The sixth episode has 7.2 million viewers. Network executives decide that they'll cancel the show if the number of viewers dips below 5 million. How many episodes will air before the show is cancelled?

I'm using the letter ______ to represent the variable quantity ____________________________.

In many applied problems, there's more than one quantity that can vary. This presents a problem when setting up an equation: if an equation has two variables, we can't solve it to get a numeric answer to our problem. In that case, we might be able to set up a **system of equations.** A system of equations is two or more related equations that have the same variables. Here's an example:

$$2x+5y=10$$
$$x-y=8$$

Problem 4 can be solved using a system of equations. Fear not – we'll guide you through the process. But first, let's set the scene.

3. The members of an intramural softball team decide to get custom t-shirts made. There are two screening shops in town that they can choose from. Wave Graphics charges a setup fee of $22, and then each shirt is $7. The Shirt Shack doesn't charge a setup fee, but each shirt is $9. If for some bizarre reason the team decided to only order 3 shirts, which shop would be the cheaper choice?

4. The goal is to find the number of shirts for which the two shops would charge the same.

a. Write an equation that describes the cost C of buying x shirts from the Shirt Shack.

b. Write an equation that describes the cost C of buying x shirts from Wave Graphics.

c. Now you have a system of equations! Good job. Let's examine the costs associated with each shop by making a table of values.

Number of shirts (x)	Cost at Shirt Shack (C)	Cost at Wave Graphics
3		
6		
9		
12		
15		

d. Based on the table, estimate the number of shirts that would make the total cost the same at each shop.

e. Graph the lines corresponding to each of the equations you wrote in parts a and b on the same coordinate system. Make sure you choose a scale so that the point where the two lines cross is visible.

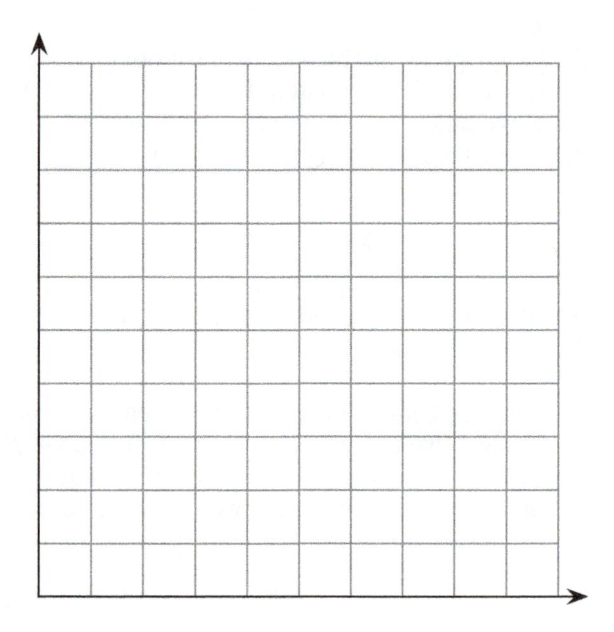

f. What quantity does the height of every point on each line represent?

g. The point where the graphs cross is where the two heights are the same. What does this represent in the problem?

h. Based on your graph, estimate the number of shirts that will lead to the same cost at each shop.

i. Our last goal is to find an exact solution by solving the system using algebra. One of the equations is $C = 9x$. This tells us that C and $9x$ are exactly the same quantity, and are interchangeable. Substitute $9x$ in for C in the OTHER equation (the one describing the cost at Wave Graphics).

j. The result of part i should be an equation with just x as variable. Solve that equation, and use the result to write a sentence describing the exact answer to the problem originally posed in Question 4.

k. What will the total cost be for the number of shirts found in part j?

Math Note

A solution to a system of equations is a *pair of numbers*, one for each variable, that makes both equations true. So after answering part k, you've solved the system.

3-7 Portfolio Name ____________________

Check each box when you've completed the task. Remember that your instructor will want you to turn in the portfolio pages you create.

Technology

1. ☐ This refers to the problem in Questions 3–11 of the Applications section of this lesson. Use your equations for the cost of buying diapers with the Amazon Mom membership and the cost without the membership to complete the table shown in an Excel spreadsheet. Then graph the system of equations. Explain how the graph shows when it is worth joining Amazon Mom.

	A	B	C
1	Boxes of diapers bought	Cost with membership	Cost without membership
2	0		
3	2		
4	4		
5	6		
6	8		
7	10		
8	12		

Skills

1. ☐ Include any written work from the online skills assignment along with any notes or questions about this lesson's content.

Applications

1. ☐ Complete the applications problems.

Reflections

Type a short answer to each question.

1. ☐ What does personal responsibility mean to you in terms of your education?
2. ☐ Describe a general approach you use when solving applied problems. Try to write it so that it would be helpful to a classmate that struggles with these problems.
3. ☐ Take another look at your answer to Question 0 at the beginning of this lesson. Would you change your answer now that you've completed the lesson? How would you summarize the topic of this lesson now?
4. ☐ What questions do you have about this lesson?

Looking Ahead

1. ☐ Read the opening paragraph in Lesson 4-1 carefully, then answer Question 0 in preparation for that lesson.

3-7 Applications

Name ______________________________

Begin by reading the article on personal responsibility linked in the online resources for Lesson 3-7.

1. List the seven choices described in the article, with a brief summary of each.

2. Are there any choices on the list that you disagree with? Explain.

3. Which of the choices are most meaningful to you? Explain.

3-7 Applications Name ______________________

Parents with young babies buy a lot of diapers (to say the very least), so getting the best deal is pretty important. Comparing prices can be harder than you would think, though – retailers don't all sell diapers in the same size boxes. A recent check of prices on Pampers Swaddlers in Size 1 revealed the following:

Amazon.com: 234 diapers for $46.99
Meijer: 160 diapers for $34.99

4. Find the unit price for each box of diapers and use it to decide which is the best deal. (There's free shipping, so don't worry about shipping or tax.)

Amazon offers an interesting deal: if you pay $79 per year to join their "Amazon Mom" club, you get 20% off all diaper purchases.

5. What would the $46.99 box of diapers cost with that discount?

6. Write an equation that provides the total cost C of buying x boxes of diapers in one year with the Amazon Mom membership.

7. Use your equation to find the cost of buying 10 boxes of diapers in a year if you join Amazon Mom.

8. Write another equation that provides the total cost C of buying x boxes of diapers in one year from Amazon without being a member of Amazon Mom.

9. Use your equation from Question 8 to find the cost of buying 10 boxes of diapers in a year without the Amazon Mom membership.

10. The equations you wrote in Questions 6 and 8 form a system of equations. Solve that system and use your result to find how many boxes of diapers you'd need to buy in a year for the two costs to be the same. What would the cost be?

11. Under what circumstances would you want to join Amazon Mom to buy diapers? (Well, obviously having a baby would be one of those circumstances. What other ones?)

Unit 4
Living in a Nonlinear World

Outline

Lesson 4-1 Is That Normal?

Learning Objectives

- ☐ 1. Compute and interpret standard deviation.
- ☐ 2. Use a normal distribution to find probabilities.
- ☐ 3. Recognize some common misuses of statistics.

There are three kinds of lies: lies, damned lies, and statistics.
– Mark Twain

In the information age, there are statistics for darn near anything you can think of, from important stuff, like mortality and employment rates, to silly stuff, like how many hot dogs the world champion can eat in 10 minutes. Even though Mark Twain lived in the 1800s, he realized that being able to understand and interpret statistics was an important way to keep from being taken advantage of. Of course, there are entire courses on statistics, and we've already talked about measures of average. In this lesson, we'll have a look at some further aspects of statistics that should help you to interpret information a little more effectively.

0. After reading the opening paragraph, what do you think the main topic of this section will be?

4-1 Group

If it doesn't make you uncomfortable, exchange the following information with the classmates in your Unit 4 group. This will be your small group for the fourth unit. It would be a good idea to schedule a time for the group to meet to go over homework, ask/answer questions, or prepare for exams. You can use this table to help schedule a mutually agreeable time.

Name	Phone Number	Email	Available times

Two golfers were competing in a televised competition for a scholarship to play on a college golf team. The scores for each golfer over all six rounds of the competition are shown in the table.

Player	Rd. 1	Rd. 2	Rd. 3	Rd. 4	Rd. 5	Rd. 6
Brittany	75	80	72	81	75	77
Ji-Min	69	77	71	75	80	83

1. Compute the mean score for each player. Round to one decimal place.

2. Find the range of scores for each golfer. (The range of a data set is calculated by subtracting the highest value minus the lowest value.)

3. Which golfer do you think was the better player in this competition? Which do you think was more consistent? Explain your reasoning.

4. Can you make up 4 exam scores that have a mean of 80, and a highest score of 100? Are the scores close together, or spread out?

5. Can you make up 4 exam scores that have a mean of 80 and a highest score of 84? Are the scores close together or spread out?

6. What do the previous two questions tell you about how there's more to the story told by a data set than just the measures of average?

4-1 Class

As we saw in the Group portion of this lesson, while the mean is a very useful tool in analyzing data, there's more to the story than just the mean because it can't factor in how spread out a group of values is. The range is one way to analyze spread, but it totally ignores all but the highest and lowest values. So if there's just one value that's really low or high compared to the others, the range will make it seem like the data are more spread out than they actually are.

For this reason, we'll study a more detailed measure of spread known as the **standard deviation.** In short, standard deviation, which is often represented by a lower-case Greek sigma (σ) is a measure of how far on average the values are from the mean. A large standard deviation means there are a lot of values far away from the mean. A small standard deviation means that most of the values are close to the mean. Here are the steps for computing standard deviation, using Brittany's golf scores from the Group portion as an example: 75, 80, 72, 81, 75, 77.

Step 1: Find the mean. Standard deviation is all about measuring distance from the mean, so we better start by finding the mean. We already found the mean in group Question 1: it's 76.7.

Step 2: Subtract the mean from each data value in the data set. This is where we're really measuring the spread.

1. Fill in the second column of the table below.

Step 3: Square the differences from Step 2. This is a simple way to eliminate the signs: we're trying to measure how far away from the mean the values are, not whether they're bigger or smaller.

2. Fill in the third column of the table by squaring each value in the second column.

Score	Score – Mean	(Score – Mean)^2
75		
80		
72		
81		
75		
77		

Step 4: Find the mean of the squares. This is almost finding the average of the distances of each data value from the mean: the only problem is that we're really averaging the squares. We'll fix that soon.

3. Find the mean of the values in the third column of the table.

Step 5: Find the square root of the mean from Step 4. In essence, this is "undoing" the squares used to eliminate the signs.

4. Find the standard deviation by computing the square root of your answer from Question 3.

Using Technology: Computing Standard Deviation

TI-84 Plus Calculator

1. Enter the data in list **L1**; press 

then 

.

2. Enter the list of values under **L1**.

3. We want choice 1 under the **STAT-CALC** menu, which you get to by pressing 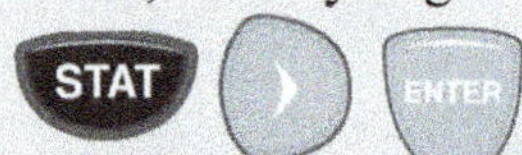

.

4. Choose **1: 1-VarStats**. Among the information displayed will be the mean, which is denoted $\bar{x}$, and the standard deviation (σx).

See the Lesson 4-1-1 and 4-1-2 videos in class resources for further instruction.

Excel Spreadsheet

B8 | f_x =STDEV.P(B2:B7)

	A	B	C
1		Score	
2		75	
3		80	
4		72	
5		81	
6		75	
7		77	
8	Standard Deviation	3.1	
9			

1. Enter the values in a row or column.
2. Enter =STDEV.P(B2: B7); note that this is the range of cells containing the data values.
3. Format the cell where you entered the formula to display the number of decimal places you want.

Note: The "=AVERAGE" and "=MEDIAN" commands in Excel are used to compute the mean and median.

The Normal Distribution

How many people do you know that are taller than 6'5", or shorter than 5'1"? For most people, the answer to that question is "not very many." The heights of humans tend to exhibit a phenomenon shared by many characteristics of living things: there are a lot of values close to the mean, and less and less as you get further from the mean. The picture to the right is an example of 100 maple seed pods that I collected from my back yard, arranged by length; note the pattern that appears.

5. Describe the pattern made by the seed pods, and explain what it says about their lengths.

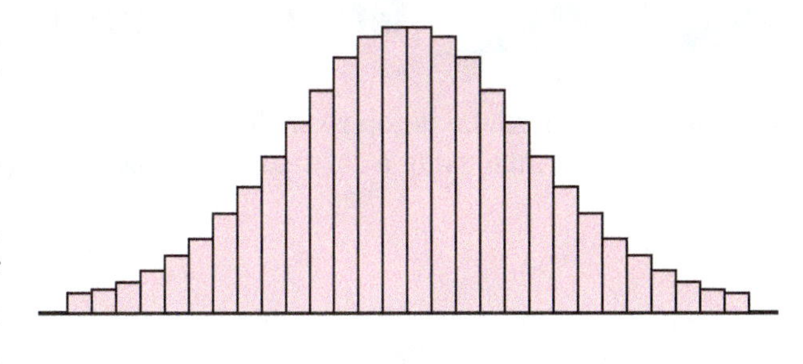

This phenomenon is so common, in fact, that data sets that follow a similar pattern are said to be **normally distributed.** Things like sizes of individuals, IQs, weights of packaged products, and lifespans of batteries or lightbulbs are often normally distributed.

I only collected 100 pods, out of thousands (maybe millions) on the tree. If I had collected a lot more of them, the picture would most likely have started to look a lot like the nice, symmetric diagram to the right. When a group of data is normally distributed, and we know the mean and standard deviation, there's a rule that allows us to estimate how many data values fall within certain ranges. This is known as the **empirical rule,** and it's illustrated by the diagram below.

The empirical rule says that when a data set is normally distributed, about 68% of all values will fall within one standard deviation of the mean; about 95% will fall within 2 standard deviations of the mean; and about 99.7% will fall within 3 standard deviations of the mean.

Math Note

The mean for data that is normally distributed is often represented by the lower case Greek letter mu (μ).

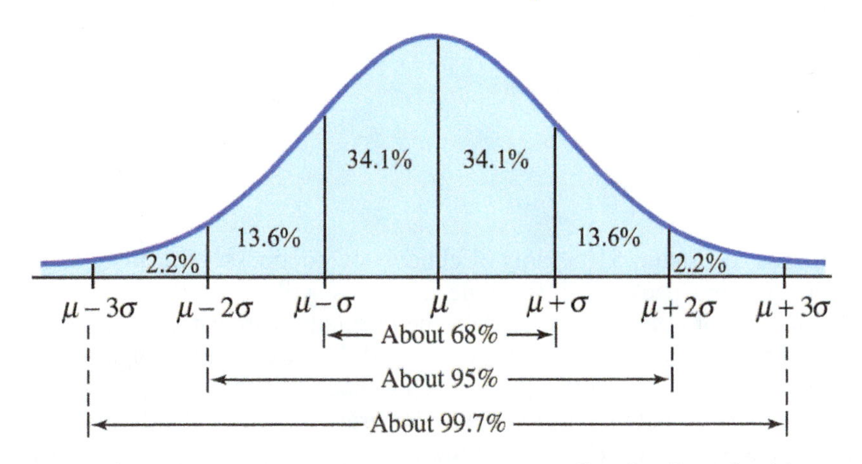

For example, the heights of American men are normally distributed with mean 5 feet 9.3 inches and standard deviation 2.8 inches. That is, μ is 5 feet 9.3 inches, and σ is 2.8 inches.

6. Use the information just given about μ and σ to fill in the blanks on the next empirical rule diagram with heights, using the formulas below the blanks for guidance.

7. Based on your diagram, about what percentage of American men fall into the height range from 5 feet 6.5 inches to 6 feet 0.1 inch?

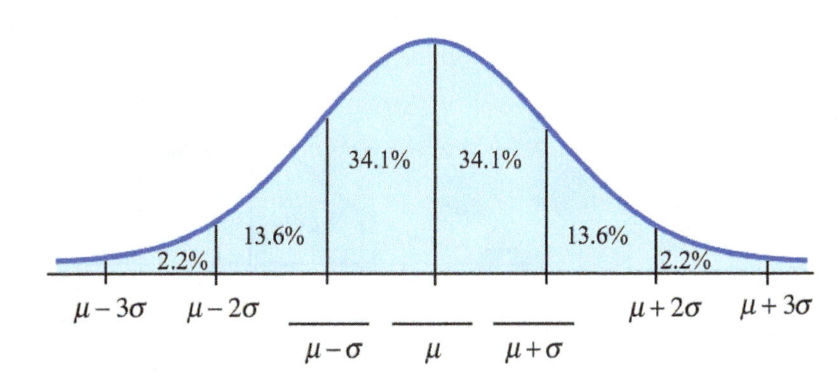

4-1 Group (Again)

1. Find the standard deviation for Ji-Min's golf scores (provided on the second page of this lesson). Compute first by hand, then using a calculator or Excel to check your answer. Round to one decimal place.

2. Discuss how comparing the two standard deviations can help you to decide which of the two golfers is a more consistent player.

A standard package of Oreos is supposed to contain 510 grams of chocolatey goodness. But there's variation in just about anything, including production and packaging, so some packages will contain more and some will contain less. In fact, this is exactly the sort of quantity that tends to be normally distributed. The folks that run Nabisco aren't stupid, and they know that customers won't be very happy if they weigh a package of cookies and find that it contains less than the labeled amount. The typical approach to keep that from happening is to design the packaging process so that the mean is something more than 510 grams, with a standard deviation that guarantees that the vast majority of packages contain 510 grams or more.

3. Let's say that the mean is 518 grams and the standard deviation is 4 grams.

   a. Fill in all of the blanks on the empirical rule diagram below with weights in grams. Use the formulas below the blanks for reference.

   b. What percent of all Oreo packages would contain between 514 and 522 grams?

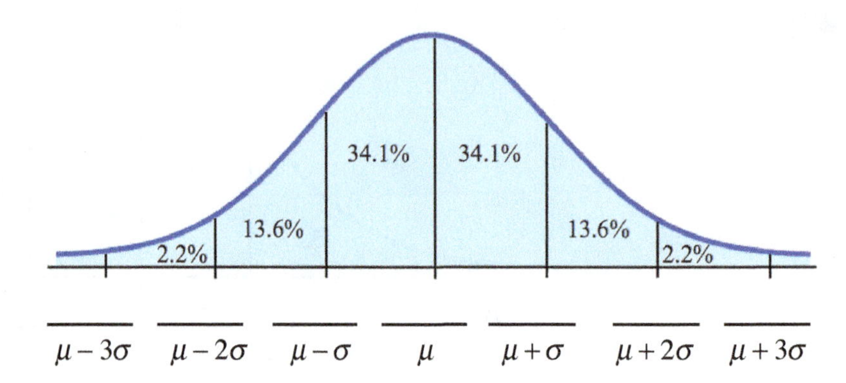

4. Using what we know about the connection between probability and percent chance, what's the probability that a randomly chosen package contains between 514 and 522 grams?

The probability of a package of Oreos containing between 514 and 522 grams can be written as $P(514 < x < 522)$, where x represents the weight. This uses a **compound inequality,** which is a combination of two inequalities. In this case, $514 < x < 522$ means that x is both greater than 514 (that's the $514 < x$ part) AND less than 522 (which is the $x < 522$ part). *In other words, x is between 514 and 522.*

5. Write a compound inequality with variable x that describes the set of all weights between 510 and 526 grams.

6. Write an expression using your answer to Question 5 that describes the probability of a randomly selected Oreo package containing between 510 and 526 grams of joy (cookies, actually).

7. Use the empirical rule to find the probability in Question 6.

8. Write a description of the probability represented by the expression $P(x < 510)$, where x represents the weight of a randomly selected package of Oreos. Shoot for a description that one of your classmates could easily understand.

9. Find the probability described in Question 8. (This will require some interpretation of the diagram illustrating the empirical rule.)

10. Based on your answer to Question 9, if 1,000 Oreo packages are sampled, how many will have less than 510 grams of cookies?

11. Is it unusual for a package to weigh more than 530 grams? Explain.

Misuses of Statistics

As our friend Mr. Twain pointed out over 100 years ago, statistics can be used to mislead if you're not careful about interpreting them. The online resources for Lesson 4-1 contain an entire supplemental section on ways that statistics can be misused, either intentionally or not. For now, we'll look at one example that's particularly relevant to the topic of this lesson.

In promotional materials, a small business claimed that their average employee makes over $150,000 per year. The table shows the annual salary for all ten employees.

Employee	Salary
William	$1,200,000
Andrew	$28,000
Marie	$42,000
Miguel	$55,000
Dwight	$60,000
Elena	$72,000
Rachel	$28,000
Kevin	$61,500
Jamaal	$40,000
Dean	$52,000

12. Is the company's claim technically truthful? Compute the mean to find out.

13. Explain why the company's claim is deceiving.

14. We studied three measures of average in Lesson 1-10. Which of those do you think would provide the most honest measure of average for this company's salaries?

4-1 Portfolio Name ______________________________

Check each box when you've completed the task. Remember that your instructor will want you to turn in the portfolio pages you create.

Technology

1. ☐ Using the Internet as a resource, find a data set that you find interesting, then use Excel to compute the mean, median, and standard deviation. Then write a brief report about what these measures of average and spread tell you about the data. Make sure that you list the source of your data.

Skills

1. ☐ Include any written work from the online skills assignment along with any notes or questions about this lesson's content.

Applications

1. ☐ Complete the applications problems.

Reflections

Type a short answer to each question.

1. ☐ What can we learn from analyzing the standard deviation for a data set that we couldn't learn from just looking at measures of average?
2. ☐ Discuss how the mean of a data set could in some cases give you a deceiving picture of what that data set is about.
3. ☐ Take another look at your answer to Question 0 at the beginning of this lesson. Would you change your answer now that you've completed the lesson? How would you summarize the topic of this lesson now?
4. ☐ What questions do you have about this lesson?

Looking Ahead

1. ☐ Read the opening paragraph in Lesson 4-2 carefully, then answer Question 0 in preparation for that lesson.

4-1 Applications Name ______________________________

1. In Lesson 1-10, you found the mean of the exam scores shown in the table. Find the standard deviation for these scores, using whatever method you prefer. Then describe what the mean and standard deviation tell you about the data.

	A	B	C
1	**Student**	**Exam 1 (%)**	**Exam 2 (%)**
2	Michael	80	89
3	Andy	77	93
4	Pam	68	84
5	Jim	81	88
6	Dwight	96	91
7	Stanley	54	75
8	Phyllis	75	54
9	Kevin	81	86
10	Creed	71	0
11	Darryl	89	83
12	Gabe	56	64
13	Toby	81	65
14	Holly	92	73

2. According to numerous online resources, the mean height for American women is 5 feet 5 inches, with a standard deviation of 3.5 inches. Use the empirical rule to find the probability that a randomly chosen American woman is between 5 feet 1.5 inches and 5 feet 8.5 inches.

3. Write an expression of the form *P*(inequality) that represents the probability you found in Question 2. Use h to represent the height of a randomly chosen woman, and write heights in inches.

4. In a group of 500 women, how many would you expect to be taller than 6 feet? (You'll need to interpret the diagram that illustrates the empirical rule.)

4-1 Applications Name ____________________

5. On one campus, about 95% of students work between 6–12 hours per week, and the number of hours worked is normally distributed. What is the mean number of hours worked likely to be? What is the standard deviation? (This one requires a little bit of ingenuity.)

6. According to the U.S. Census Bureau, the mean price of all new homes sold in the United States in April of 2013 was $330,800. The median price, on the other hand, was $271,600.

a. What do you think accounts for the large discrepancy in these two measures of average?

b. Discuss which of the two measures you think is a more accurate representation of what the typical home buyer could expect to pay for a new home.

Lesson 4-2 A Road Map to Success

Learning Objectives

- ☐ 1. Understand and use the Pythagorean theorem.
- ☐ 2. Read contour maps and calculate grade.
- ☐ 3. Develop and use the distance formula.

Map out your future, but do it in pencil. The road ahead is as long as you make it.
– Jon Bon Jovi

While a road map to success is a useful metaphor, in this section we'll study a particular type of map that will help us to introduce some key ideas in geometry. When you're looking at a road map, it's very much like looking straight down from a plane, and the roads all look two-dimensional. But of course, we live in a three-dimensional world, and what you're missing is the change in height. Because of this, the actual distance you'd drive on a road isn't always the distance you'd measure on a map. We'll use the change in height for roads to study two important results used to study our physical world: the Pythagorean theorem and the distance formula.

0. After reading the opening paragraph, what do you think the main topic of this section will be?

4-2 Class

The road sign above describes a mountain road with a grade of 6%. This is a warning that the road is steep enough to require some extra caution, but what exactly does it mean? The slope of a road is very much like the slope of a line: it's a comparison between how far the road goes up or down for a certain horizontal distance. A grade of 6% means that for any horizontal distance, the change in height is 6% of that distance.

1. What's the change in elevation over a 2,000 foot horizontal stretch for a road with a 6% grade? Fill in that change in height on the diagram.

2,000 ft

The diagram now illustrates the situation we described in the lesson opener: the horizontal distance (2,000 feet) would appear on a map of the road, but the slanted line at the top of the triangle represents the road. So the actual distance you drive is the length of that slanted side, which we don't know based on the horizontal distance and the change in height.

Fortunately, there's a very old and very famous result from geometry that will allow us to find the actual driving distance: the Pythagorean theorem. This provides a mathematical relationship between the lengths of sides in a triangle when two of the sides are perpendicular. In the diagram below, we'll call the lengths of the three sides *a*, *b*, and *c*. The little square at one of the angles indicates that those two sides are perpendicular: we call that angle a **right angle,** and the triangle a **right triangle.**

The Pythagorean Theorem

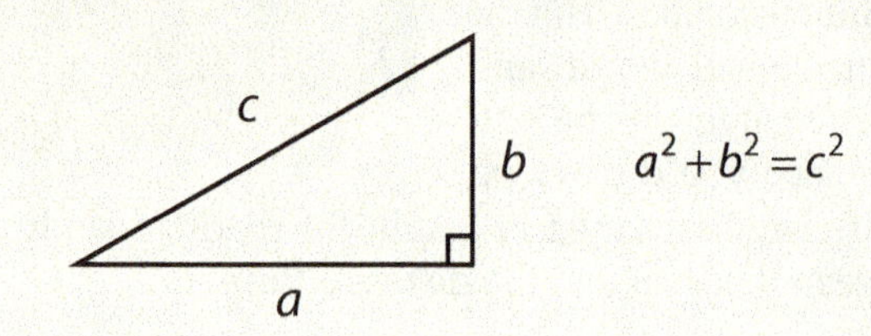

$$a^2 + b^2 = c^2$$

In words, the Pythagorean theorem says that if you square the lengths of the two shorter sides and add the results, you'll get the square of the length of the longest side. The longest side is called the **hypotenuse,** and it's always across from the right angle. Let's look at how we can use the Pythagorean theorem to study grade.

2. Notice that the diagram at the bottom of page 245 is a right triangle. Use the Pythagorean theorem to write an equation involving the length of the missing side of the triangle. Then simplify the side with only numbers on it.

3. In order to solve your equation to find the missing length, you'll need to "undo" the square: this is exactly what square roots are designed to do. So apply a square root to both sides to find the missing length.

4. How much farther would you drive on that road than the 2,000 feet that would appear on a map?

Math Note

There are actually two numbers that make the equation $c^2 = 4{,}014{,}400$ true: 2,003.6 and −2,003.6. But in this setting, *c* is a physical distance, so we ignore the negative solution in Question 3.

4-2 Group

1. Suppose that we know the distance between points A and B is exactly five miles. If the contour of the road followed the solid path in each of the three figures below, would the distance traveled by the car be more than, less than, or equal to five miles?

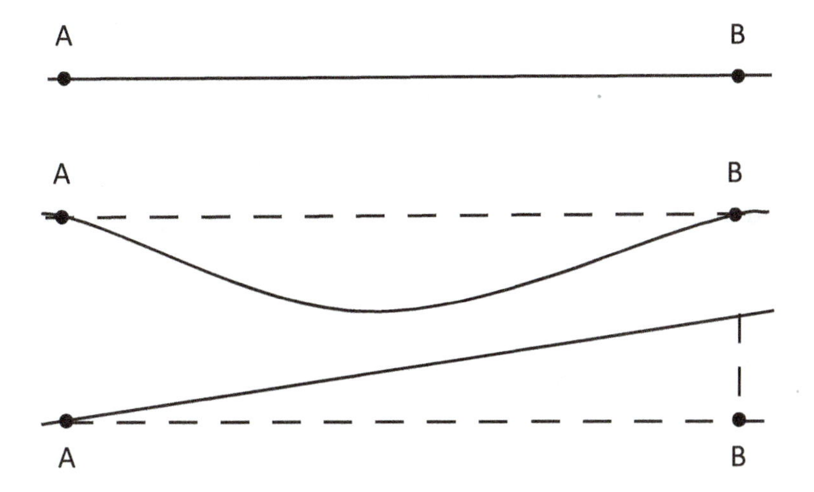

Contour maps might remind you of the isotherm maps we learned about back in Lesson 1-4. But instead of the lines representing locations that have the same temperature, the lines represent locations that have the same altitude. This is a way to overcome the two-dimensional nature of maps, allowing us to see both straight-line distance from above, and changes in height as well. The top part of the next diagram is the contour map; under the map is an illustration of a side view that shows what the elevations look like based on the map.

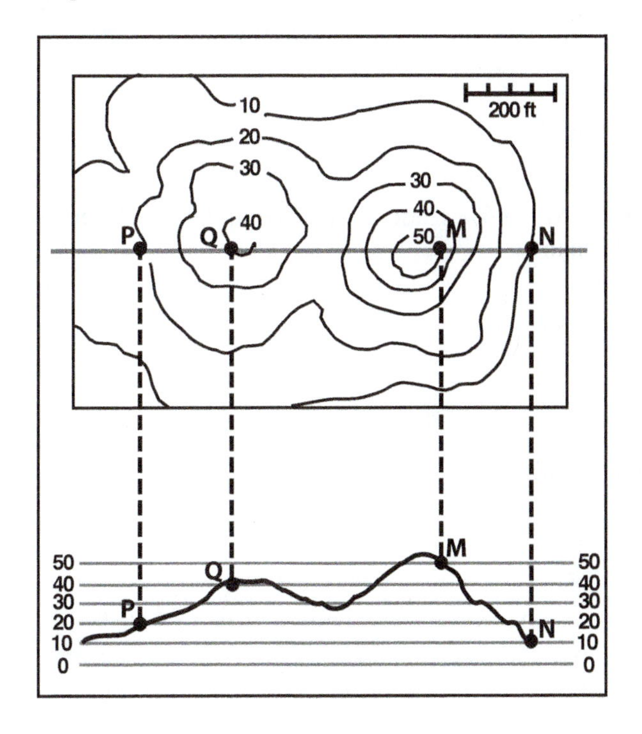

2. How does the distance between points P and Q compare to the distance between points M and N on the map? Use the scale on the map.

3. The lower illustration also shows points P, Q, M, and N, but this time the elevations are shown as well. How does the actual hiking distance from P to Q compare to the hiking distance from M to N? What can you conclude about apparent distance on a contour map?

4. Based on the scale provided on the two-dimensional contour map, the distances from P to Q and from M to N are both about 200 feet. Use the lower illustration to estimate the actual trail length between points P and Q, making sure to describe how you got that estimate. Then repeat for the trail between points M and N.

5. Illustrate the grade of each trail by drawing a right triangle.

6. Use the Pythagorean theorem to estimate the true length of each trail.

7. The road sign at the beginning of this lesson is from a road with a 6% grade for a horizontal span of 8 miles. By how much does the elevation change over that 8 mile span?

8. How far would you actually drive in covering that 8 mile span?

9. Before selling his house, Brian needs to remove a dead mouse that got trapped under the bathtub. (Clearly, "includes a deceased animal" is not a feature buyers are looking for.) Using the corner of the room as the origin, and using a patented, high-tech rodent corpse location device, he was able to determine that the mouse was located at coordinates (4, –2). An access panel under the sink is located at coordinates (1, –5). Assuming that the units are feet, the goal is to find the distance from the access panel to the mouse.

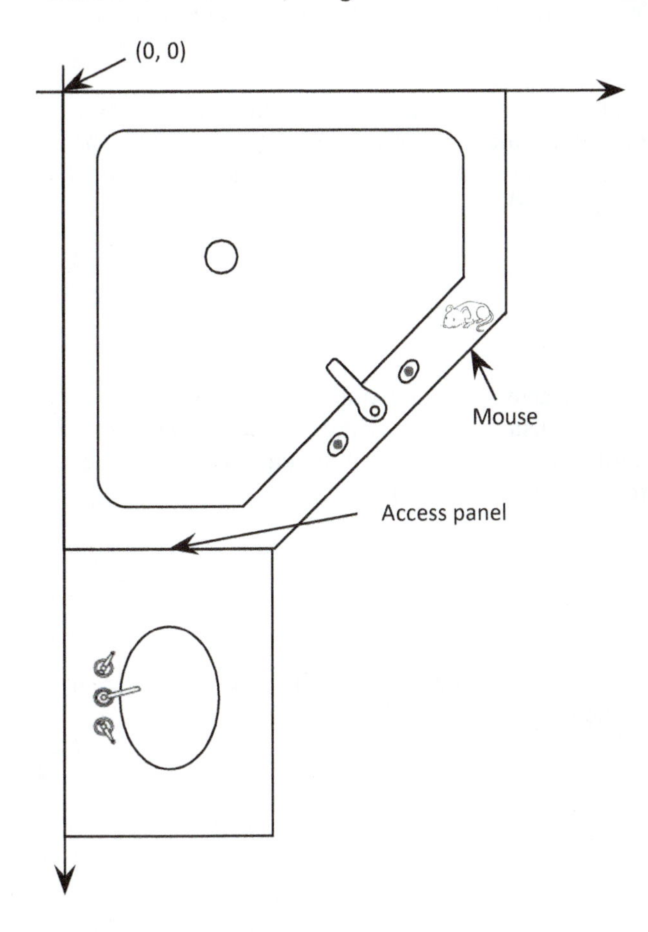

10. Estimate the distance by plotting the two given points on graph paper, then using a second sheet of graph paper as a ruler.

11. Using your plot from Question 10, use the Pythagorean theorem to find the exact distance, first in exact form (which will have a square root in it), then in decimal form rounded to the nearest tenth of a foot.

4-2 Class (Again)

The process used to find the distance from the access panel to the mouse in Group Questions 9-11 can be mimicked to develop a generic formula for finding the distance between any two points on a graph. Let's start with two arbitrary points, labeled (x_1, y_1) and (x_2, y_2). Since we're not labeling specific number coordinates, but using symbols, these points can represent ANY pair of points.

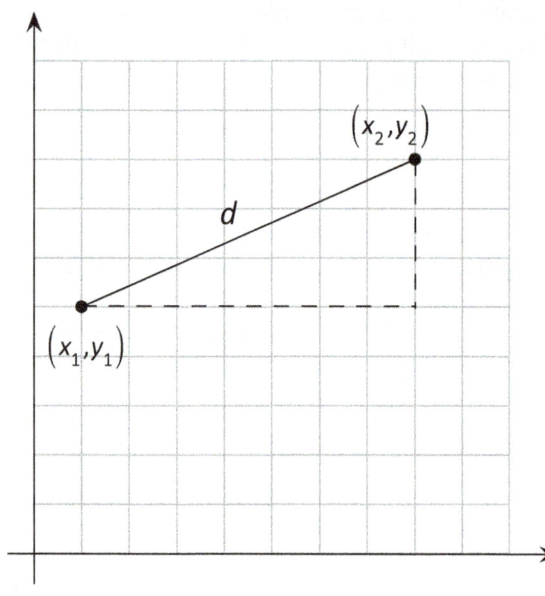

1. What are the coordinates of the point where the two dotted lines meet to form a right angle? (Hint: That point is at the same height as (x_1, y_1) and lives directly below (x_2, y_2).)

2. Use your answer to Question 1 to find the length of the two dotted sides of the right triangle, then label those lengths on the diagram. Taking another look at the mouse problem calculation might help.

3. Use the Pythagorean theorem to set up an equation containing the distance d that we're looking for. (Hint: Putting parentheses around the expression you wrote for the lengths in Question 2 is a good idea.)

Math Note

Even though the diagram we're using is in the first quadrant, the fact that we used arbitrary points means that our formula will work for any two points.

4. Solve the equation to find a formula for the distance between any two points.

5. A map of a national park has a grid on it to help reference locations. The camping office is located at the point (–4, 10) on the grid, and the lodge is at (2,4). Each square on the grid represents a distance of two miles. Use your formula from Question 4 to find the distance from the camping office to the lodge.

4-2 Portfolio

Name ______________________

Check each box when you've completed the task. Remember that your instructor will want you to turn in the portfolio pages you create.

Technology

1. ☐ Search online for a contour map of a place that you'd like to visit someday. Copy and paste the map into a Word document, then write a description of what you can learn about that place from reading the contour map.
2. ☐ Research the history of the Pythagorean theorem on the Internet and draw a timeline illustrating your findings.

Skills

1. ☐ Include any written work from the online skills assignment along with any notes or questions about this lesson's content.

Applications

1. ☐ Complete the applications problems.

Reflections

Type a short answer to each question.

1. ☐ What does the Pythagorean theorem say? When can you use it?
2. ☐ Why can't you just find the distance between two points by plotting them on a grid, then counting the number of boxes between them?
3. ☐ Take another look at your answer to Question 0 at the beginning of this lesson. Would you change your answer now that you've completed the lesson? How would you summarize the topic of this lesson now?
4. ☐ What questions do you have about this lesson?

Looking Ahead

1. ☐ Read the opening paragraph in Lesson 4-3 carefully, then answer Question 0 in preparation for that lesson.

4-2 Applications Name ____________________

Questions 1-3 are based on the contour map provided, which is of Mt. St. Helens in Washington. All elevations on the contour lines are given in meters.

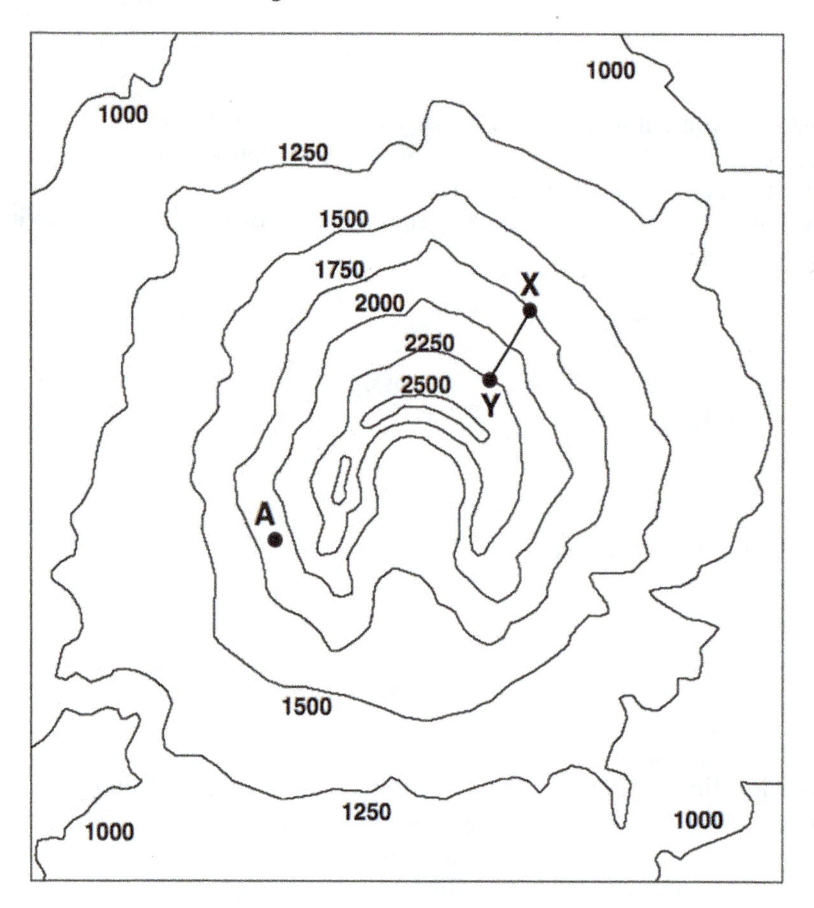

1. Estimate the elevation of point A.

2. If points X and Y are 4,000 m apart looking straight down at the map, find the grade of a trail that connects those two points. (Hint: Find the change in elevation first.)

3. How far would you walk along the trail from point X to point Y? Assume that the elevation changes in a straight line. Round to the nearest tenth of a meter, then convert the distance to feet and miles.

4. An archaeological dig is marked with a rectangular grid where each square is 4 feet on a side. An important artifact is discovered at the point corresponding to (–50, 25) on the grid. How far is this from the control tent, which is at the point (20, 30)?

Lesson 4-3 Irate Ducks

Learning Objectives

- ☐ 1. Recognize when a graph is parabolic.
- ☐ 2. Solve problems using the graph of a quadratic equation.

The moral arc of the universe bends at the elbow of justice.

– Martin Luther King, Jr.

All of the graphs we studied in Unit 3 have one thing in common: they don't change direction. As Dr. King's quote implies, the term "arc" is often used to describe a graph that's not a straight line, the key feature being change in direction. One of the simplest curved graphs we can study is also very useful in modeling things in our world, as we will see very quickly in studying some very angry ducks. This ties in well to the theme of this entire unit: While many things can be modeled well with linear equations, there are many things in our world that aren't linear. In Lesson 4-2 we looked briefly at trails and roads that don't always follow a straight-line path. In this lesson, we'll begin a more detailed look at our nonlinear world.

0. After reading the opening paragraph, what do you think the main topic of this section will be?

4-3 Class

1. Ducks seem awful mild-mannered, always floating placidly on a pond, or waddling in a little line across the street. But like anyone else, push them too far and it turns out that they can be downright irate! (Especially when you don't have legal permission to use a different group of angry birds.) A group of beavers has started to build a dam that will flood our ducks' nests, and it's time to strike back! Draw four flight paths: one from the duck cannon on the left to each of the beavers on the right. Keep in mind that gravity exists! It's not possible to fly in a straight-line path. And birds can't fly through wooden posts. Use your imagination to change the launch angle of the cannon as necessary.

2. Describe the shape of each path in your own words. How are they similar? How are they different?

The shape of the path that an object moving through the air under the influence of gravity follows is called a **parabola.** This is illustrated in the time-lapse photo of a bouncing ball. Parabolas have several distinguishing features, but the two most obvious are the change of direction when reaching a high (or in some cases low) point, and the fact that the two halves on either side of the direction change are mirror images. This is known as **symmetry.**

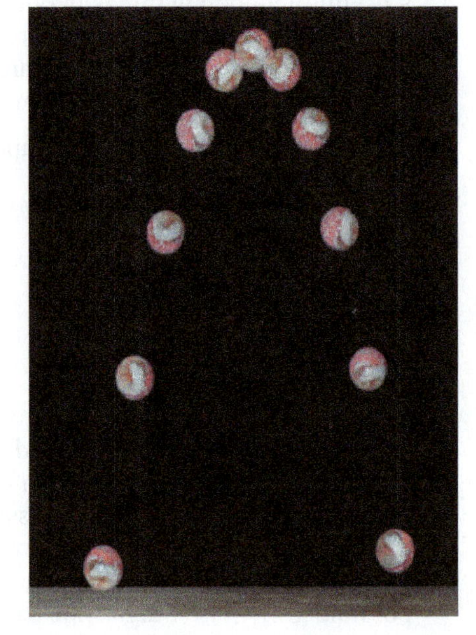

3. The amount of profit that a company makes when they produce x units of a product can often be represented by a parabola similar to the one in the photo. Of course, if a company makes no product they'll lose money. Explain why a parabola is a good choice to model the profit. (Hint: How much profit will they make if they produce no items? What if they produce so many they can't sell them all?)

4-3 Group

A homeowner is planning to fence in a play area for his dog Moose, and he decides that he'd like to spend no more than the cost of 120 feet of fence. But since he wants Moose to have plenty of room for chasing squirrels, he decides to use the full 120 feet. He also wants the play area to be rectangular, as shown in the diagram. **Answer each question (except 3 and 8) with a full sentence.**

1. If he decides to make the play area 5 feet wide, how long would it be? Don't forget to consider all four sides.

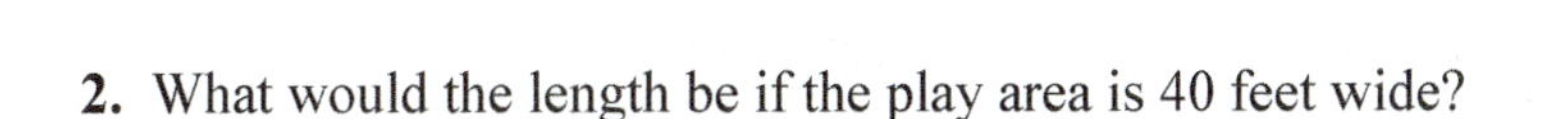

2. What would the length be if the play area is 40 feet wide?

3. Complete the table, then plot the points; your ordered pairs should look like (width, area), or (w, A). Recall that the area of a rectangle is the length times the width.

Width	Length	Area	Ordered pair (w, A)
5			
10			
15			
20			
25			
30			
35			
40			
45			
50			

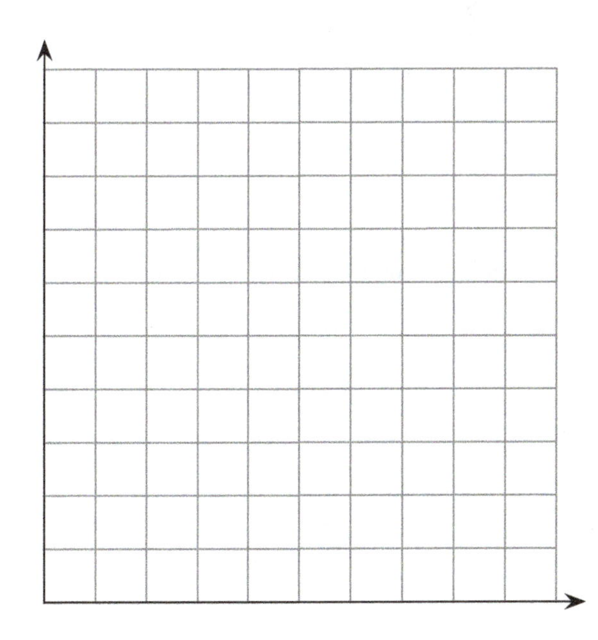

4. Based on the graph, what width makes the area as large as possible?

5. What's the area of the largest possible pen he can make out of 120 feet of fence?

6. Describe the shape that the play area should be to make the area as big as possible.

7. Write an explanation of the process used to calculate the length needed for each width.

8. Write an expression with w representing the width to describe the length when the width is w feet. Your answer to Question 7 should be a big help.

9. Multiply the width w by your expression for length from Question 8 to get an expression for area. If you multiply out any parentheses, what's the largest power of w that appears?

The profit made by a recycling business is illustrated by the following graph. The first coordinates represent the number of tons of recycled material they take in during a month. We'll use the letter x to represent this variable quantity. The second coordinates of points are the company's profit in dollars. **Answer every question with a complete sentence.**

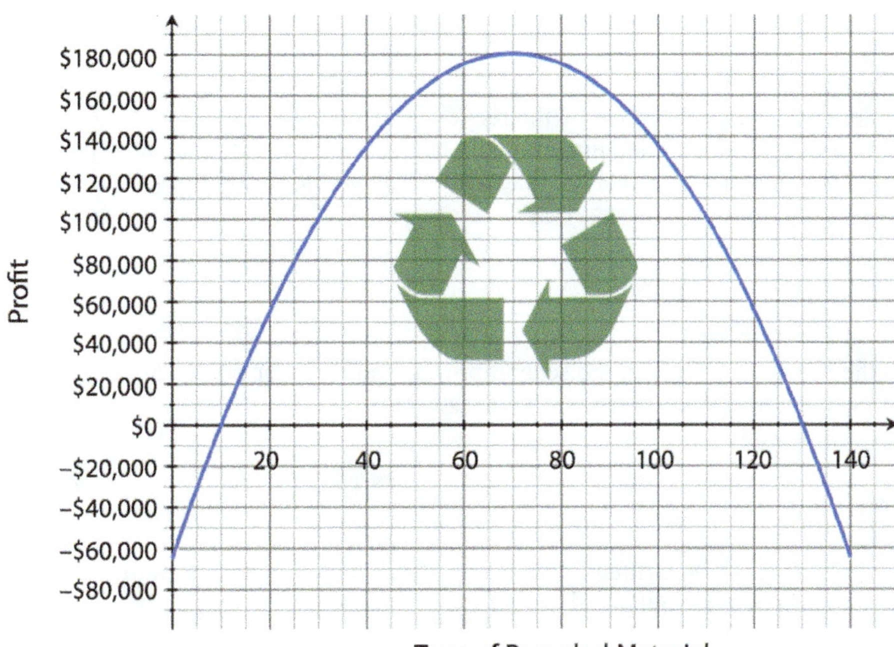

10. What's the y intercept of the graph? What does it tell us?

11. What are the x intercepts of the graph? What do they tell us?

12. How many tons of recycled material will they need to take in in order to make a profit of $100,000?

13. What's the maximum profit the business can generate? How many tons of recycled material are needed to reach that profit?

14. The equation that describes the profit made by the recycling company is $y = -50x^2 + 7{,}000x - 65{,}000$, where x represents the number of tons of recycled material they take in, and y is the associated profit. Use a graphing calculator or computer program to recreate the graph from this problem, and use a trace and/or a table feature to verify your answers to Question 13.

An equation in which the variable appears to the second power, and no other power except perhaps first, is called a **quadratic equation.** Here's the general form for a quadratic equation:

Definition of a Quadratic Equation

An equation is quadratic if it can be written in the form $y = ax^2 + bx + c$, where all of a, b, and c are numbers. Note that a can't be zero, but b and c can be.

The fact that a can't be zero in the definition of a quadratic, but b and c can be zero means that to qualify as quadratic, an equation has to have a term with the variable squared, and it may or may not also have a term with the variable to the first power, and a numeric term. Notice also that the definition says "can be written in the form". Sometimes you may have to do a bit of simplifying to recognize a quadratic equation.

15. Which of the following equation is quadratic?

a. $y = 4x^2 + 16x - 3$

b. $y = 8x - x^2$

c. $y = x^2 + 2x^3$

d. $y = 2x(x - 4)$

e. $y = 2x^2 + 3\sqrt{x}$

Math Note

Other than the form of the equation, every quadratic equation has one main thing in common: the graph is a parabola.

4-3 Portfolio

Name ____________________

Check each box when you've completed the task. Remember that your instructor will want you to turn in the portfolio pages you create.

Technology

1. ☐ Use Excel to create a graph describing the width and area of the rectangular play area in the Group portion of this lesson. Use the table on Page 255 as a model for your spreadsheet. A smooth marked scatter plot should get the job done nicely. Print your graph and add labels on the key points that were discussed in Questions 1, 2, 4, and 5.

Skills

1. ☐ Include any written work from the online skills assignment along with any notes or questions about this lesson's content.

Applications

1. ☐ Complete the applications problems.

Reflections

Type a short answer to each question.

1. ☐ If you're looking at a graph, what are some key things that will tell you that it might be a parabola?
2. ☐ How can you tell if an equation is quadratic?
3. ☐ Think of some quantities that could be modeled with quadratic graphs.
4. ☐ Take another look at your answer to Question 0 at the beginning of this lesson. Would you change your answer now that you've completed the lesson? How would you summarize the topic of this lesson now?
5. ☐ What questions do you have about this lesson?

Looking Ahead

1. ☐ Read the opening paragraph in Lesson 4-4 carefully, then answer Question 0 in preparation for that lesson.

4-3 Applications Name ____________________

After a football is kicked by a punter, it's subjected to gravity, and follows a parabolic path. The height of one particular punt is shown in the graph. The first coordinates represent the horizontal distance from the punter, and the second coordinates represent the height of the ball above the ground. Both measurements are in feet. **Answer each question with a full sentence.**

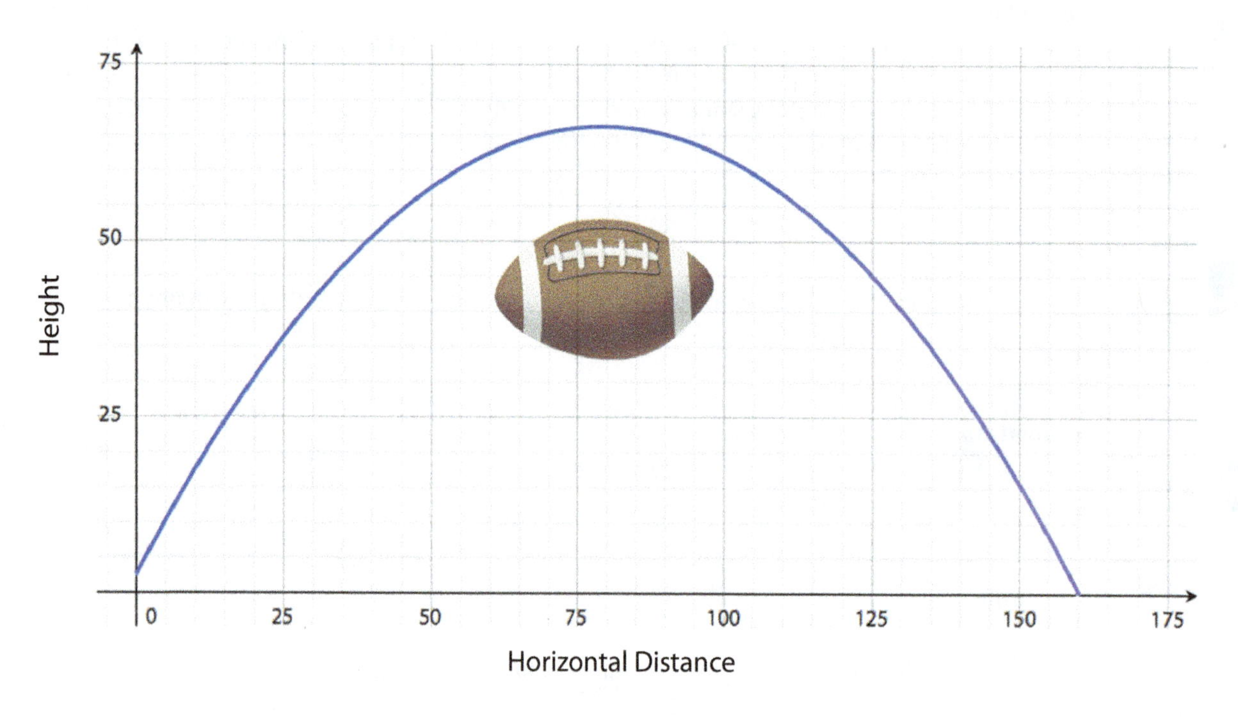

1. About how high above the ground was the ball when it was kicked?

2. How far away from where it was kicked did the ball land?

3. How high did the punt go?

4. Use the graph and some ingenuity to estimate the actual distance the punt traveled. (This is different than how far away it landed! Think of the parabola as a road, and estimate how far you'd travel on that road.) There's no right answer here: a big part of the question is describing how you got your estimate.

4-3 Applications Name ____________________

The next graph also shows the height of the ball on the y axis, but this time the x coordinates are the number of seconds since the ball was kicked.

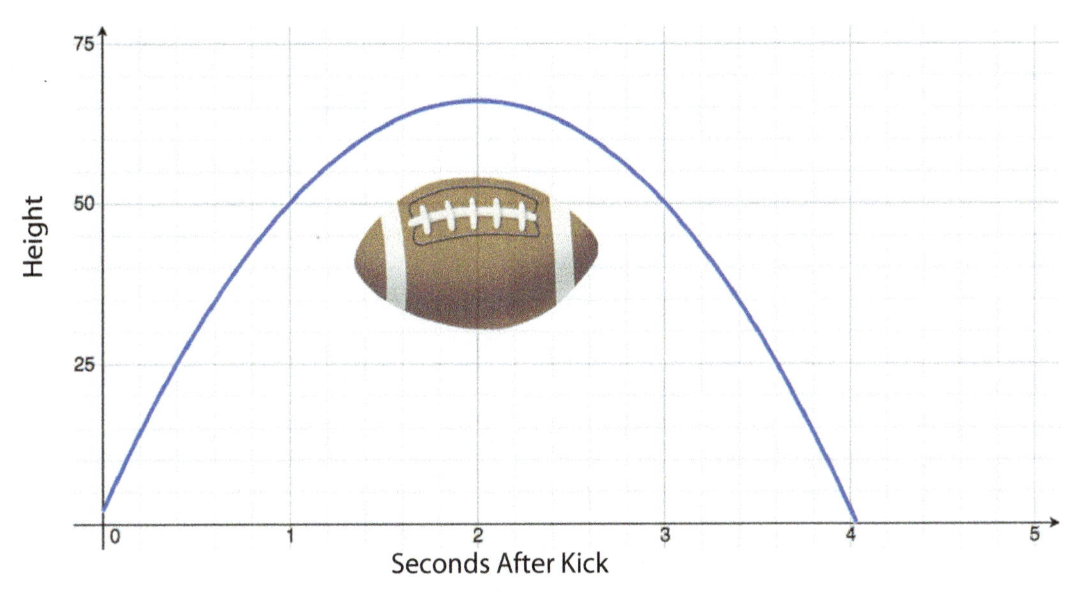

5. The hang time of a punt is how long it's in the air before returning to the ground. What was the hang time for this punt?

6. After how many seconds was the ball 60 feet above the ground?

7. The equation of the parabola representing the ball's height is $y = -16(x-2)^2 + 66$, where y is the height in feet and x is the number of seconds after it was kicked. Find the exact height after the number of seconds you found in Question 6. How accurate were your answers to Question 6?

8. If you find the value of y when $x = 5$ in that equation, why does the result not apply to the punt?

9. Explain why the shape of the graph on the previous page does NOT tell you what the flight path of the ball looked like from a side view.

Lesson 4-4 Sit Back and Watch Your Money Grow

Learning Objectives

- ☐ 1. Revisit exponential growth.
- ☐ 2. Solve problems using graphs representing exponential growth and decay.

Exponential growth looks like nothing is happening, and then suddenly you get this explosion at the end.
– Ray Kurzweil

Raise your hand if you hope to be rich someday. There's a pretty good chance that your hand is up right now: financial independence is a goal for a lot of people. We've already studied the differences between quantities that grow linearly and those that grow exponentially. Now that our skills at interpreting graphs have grown, well, exponentially, it's a good time to revisit exponential growth to learn a little bit more about its magic. We already know that the long-term growth of money is an ideal way to study exponential growth, so we'll start there, and on the way start you on the path to riches!

0. After reading the opening paragraph, what do you think the main topic of this section will be?

4-4 Class

1. Describe the differences between linear and exponential growth based on what we know from earlier in the course.

2. What do you think is meant by the "relative change" in a quantity? How do you compute relative change? (Referring to the topic of relative difference in Lesson 2-4 should help.)

If you're like a lot of people, you'd go on a spending spree if you won $10,000 on a game show. But what if you decided to invest that money rather than spend it, and what if you were patient enough to let it grow for 40 years in a savings account that compounds interest annually? Let's take a look. Make sure that you answer every question with a complete sentence.

Time (years)	Value ($)
0	$10,000.00
1	$10,700.00
2	$11,449.00
3	$12,250.43
4	$13,107.96
5	$14,025.52
6	$15,007.30
7	$16,057.81

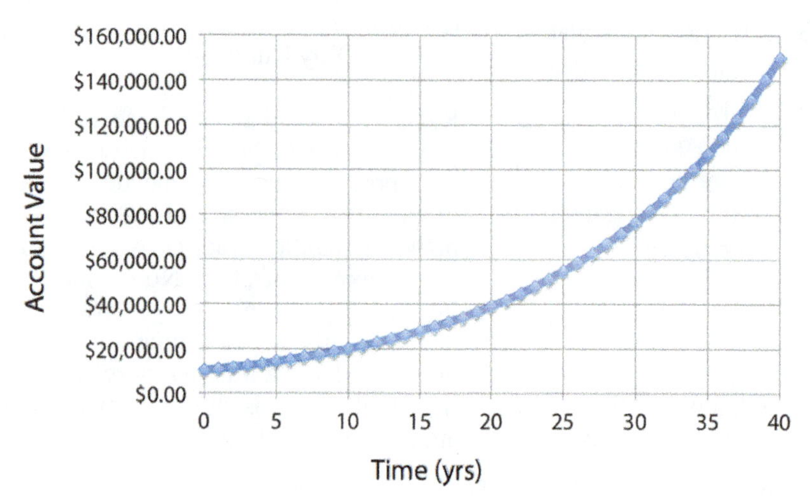

3. Keep in mind that the percent interest earned remains constant for the life of the account. What is the interest rate earned each year? How can you tell?

4. Find the relative change in value for each year in the table compared to the previous year. What can you conclude?

5. What's the multiplication factor that you'd need to multiply by a previous year's value to get the next year's value? (Hint: If my account gained 6% of its value, I'd have 106% of the amount from the previous year.)

6. Use your answer to Question 5 to write an equation in the form "$y =$ expression," where the expression describes the amount of money in the account after x years. (Hint: Starting with $10,000, how many times have you multiplied by your answer to Question 5 after x years?)

7. Use a graphing calculator to verify that the equation you wrote in Question 6 gives you the values in the table, and the given graph. Describe your results.

8. How much interest will you earn from zero to five years into the investment? This will be the difference in the value after five years and the initial investment. First, estimate the answer by looking at the graph. Then perform a calculation using the table provided.

9. How much interest will you earn between years 35 and 40? Again, estimate from the graph first. Then use the equation you wrote to get an exact value.

10. Describe what you can learn from comparing the answers to Questions 8 and 9.

4-4 Group

We've now studied exponential growth several times in this course. But there's a flip side to exponential growth, known as exponential decay. The thing that distinguishes exponential growth is that the growth starts off relatively slowly, then speeds up dramatically as time passes. Let's see how exponential decay compares.

Coffee is a popular morning drink because the caffeine it contains acts as a temporary stimulant. Once the caffeine gets into your system, the amount dissipates exponentially (assuming that you don't add more, of course). Answer every question (except for 2 and 6) with a full sentence.

Time (hours)	Caffeine (mg)
0	180.0
1	144.0
2	115.2
3	92.2
4	73.7
5	59.0
6	47.2
7	37.7

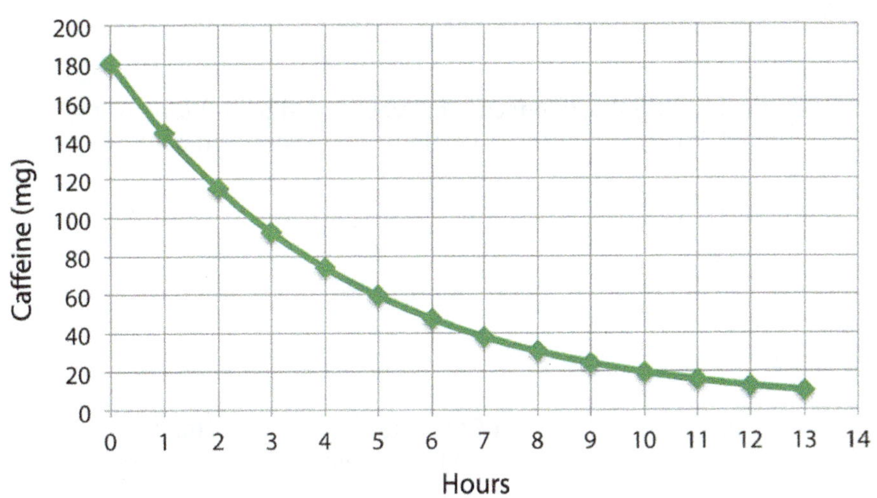

1. If you're the coffee drinker represented by the table and graph, how much caffeine was initially in your system?

2. Find the relative change in caffeine for each one-hour period listed in the table below. Round to the nearest full percent.

Hours	Relative change
0-1	
1-2	
2-3	
3-4	
4-5	
5-6	
6-7	

3. Complete this important statement about exponential change: A quantity either grows or decays exponentially when the _______________ _______________ in that quantity is _______________.

4. Looking at the graph of the caffeine remaining in your system, describe what exponential decay looks like compared to exponential growth.

5. What's the multiplication factor that you'd need to multiply by a previous hour's amount to get the next hour's amount? (Hint: If 10% of something goes away each hour, 90% remains.)

6. Use your answer to Question 5 to write an equation of the form "y = expression" where the expression describes the amount of caffeine left in your system after x hours. (See hint for Question 6 on page 264.)

7. Use a graphing calculator to verify that the equation you wrote in Question 6 gives you the values in the table, and the given graph. Describe your results.

8. Use the TRACE feature on your graphing calculator to find the amount of caffeine left in your system after 90 minutes. (Pay attention to units!)

9. Use the graph to estimate the amount of time needed for the amount of caffeine to drop to 40 mg.

10. Use either TABLE or TRACE commands on your calculator to try and come up with a more accurate estimate of the time needed for the caffeine level to reach 40 mg.

11. Make up a situation that could be modeled by each of the following exponential graphs. As always, extra points for creativity!

a.

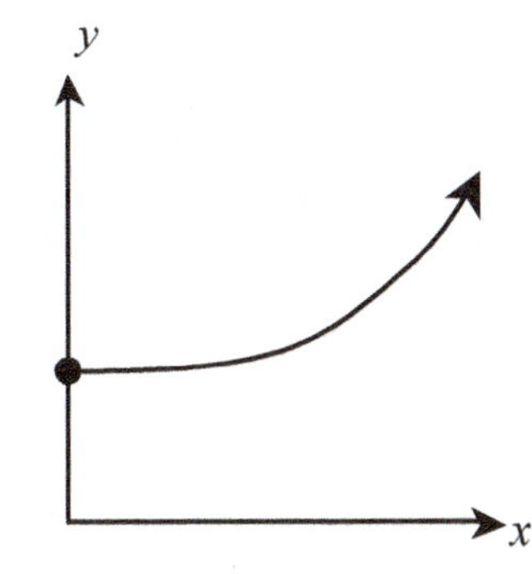

b.

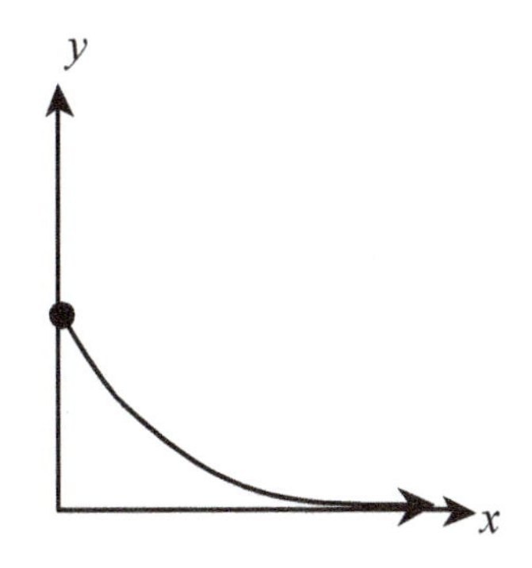

Math Note

The relative change for a quantity that grows or decays exponentially is called its **relative growth rate** or **relative rate of decay.**

4-4 Portfolio

Name ____________________

Check each box when you've completed the task. Remember that your instructor will want you to turn in the portfolio pages you create.

Technology

1. ☐ Use an Excel spreadsheet to extend both the table and the graph for the caffeine problem (p. 267) out to the point where there's less than 1 mg left in your system. How long does this take? A template to help you get started can be found in the online resources for this lesson.

Skills

1. ☐ Include any written work from the online skills assignment along with any notes or questions about this lesson's content.

Applications

1. ☐ Complete the applications problems.

Reflections

Type a short answer to each question.

1. ☐ Compare exponential growth and exponential decay. What do they have in common? How are they different? A really good answer will say something about relative change.
2. ☐ What did you learn in this lesson about the value of leaving money invested for long periods of time? How can you apply this to retirement?
3. ☐ Take another look at your answer to Question 0 at the beginning of this lesson. Would you change your answer now that you've completed the lesson? How would you summarize the topic of this lesson now?
4. ☐ What questions do you have about this lesson?

Looking Ahead

1. ☐ Read the opening paragraph in Lesson 4-5 carefully, then answer Question 0 in preparation for that lesson.

4-4 Applications Name ____________________

This Applications activity is based on an article about the growth of retirement savings, which is available in the online resources for this section. Download and read the article, then use the information to answer each question.

1. How much money is Joe earning when he's 30?

2. How much money is he allowing himself to spend when he's 30? Show a calculation, then describe whether or not your answer seems correct based on the graph in the article.

3. How much is Joe earning when he's 35? Show a calculation based on the annual relative growth rate of 4%. Then describe whether or not your answer seems correct based on the graph.

4. How much money is Joe allowing himself to spend when he's 35? Show a calculation, then describe whether or not your answer seems correct based on the graph in the article.

5. Carefully explain the meaning of the crossover point.

6. How much less money does Fran live on when she's 30 than Joe does? How much earlier does this allow her to reach her crossover point?

4-4 Applications Name ______________________

7. Use the first graph in the article to estimate each amount:

a. The amount Joe earns from his job when he's 60.

b. The amount Joe allows himself to spend when he's 60.

c. The amount Joe earns from his investments when he's 60.

8. Write an equation of the form "y = expression" where the expression describes the amount Joe earns x years after age 30.

9. Use your answer to Question 8 to find how much Joe earns at age 60. How does it compare to the estimate you got from the graph?

10. Write an equation of the form "y = expression" where the expression describes the amount Joe spends x years after age 30.

11. Use your answer to Question 10 to find how much Joe spends at age 60. How does it compare to the estimate you got from the graph?

12. Describe what impact this article might have on your saving vs. spending plan in the future.

Lesson 4-5 Where's My Jetpack?

Learning Objectives

- ☐ 1. Recognize inverse variation.
- ☐ 2. Solve problems involving direct and inverse variation.

Nothing is too small to know, and nothing is too big to attempt.

– William Cornelius Van Horne

If you talk about modes of travel to most Americans that grew up before the 1990s, you'll likely find that they feel kind of cheated. Based on movies, TV shows, and books from the 1950s through the 1980s, we all pretty much figured we'd have flying cars and jetpacks by the time 2013 rolled around, just like James Bond. What a letdown! For the most part, we're still stuck with modes of transportation that have been around for over 100 years. Sigh. Jetpacks exist of course, but they're not exactly widely available. In this lesson, we'll use modes of transportation, including the long-promised jet pack, to study the relationship between speed and time for a given trip. Clearly those quantities are related, but it turns out that exactly how they're related will lead us to study a new type of variation.

0. After reading the opening paragraph, what do you think the main topic of this section will be?

4-5 Group

In studying direct variation in Lesson 3-4, we found that when two quantities vary directly, if one goes up the other does as well. But think about speed and time when traveling: the faster you go, the sooner you get there. So as speed goes up, time goes down. The goal of this activity is to study that relationship in depth. Let's say you need to travel 40 miles, and you have several different choices of how to get there.

1. For each mode of travel, find how long it will take to make the 40 mile trip.

Mode	Speed (mi/hr)	Time (hr)
Mosey	1	
Walk	3	
Jog	5	
Bike	10	
Bus	20	
Car	40	
Jetpack	80	

2. In going from jogging (5 mi/hr) to biking (10 mi/hr), the speed doubles. What happens to the time?

3. In going from bus (20 mi/hr) to car (40 mi/hr), the speed doubles. What happens to the time?

4. In going from bus (20 mi/hr) to bike (10 mi/hr), the speed is cut in half. What happens to the time?

5. In going from jetpack (a zippy 80 mi/hr) to bus (20 mi/hr), the speed is divided by four. What happens to the time?

6. Does time depend on speed in this problem, or does speed depend on time? Explain your answer.

7. Graph the values in your table and connect them with a curve. Put the independent variable (as decided in your answer to Question 6) on the x axis and the dependent variable on the y axis.

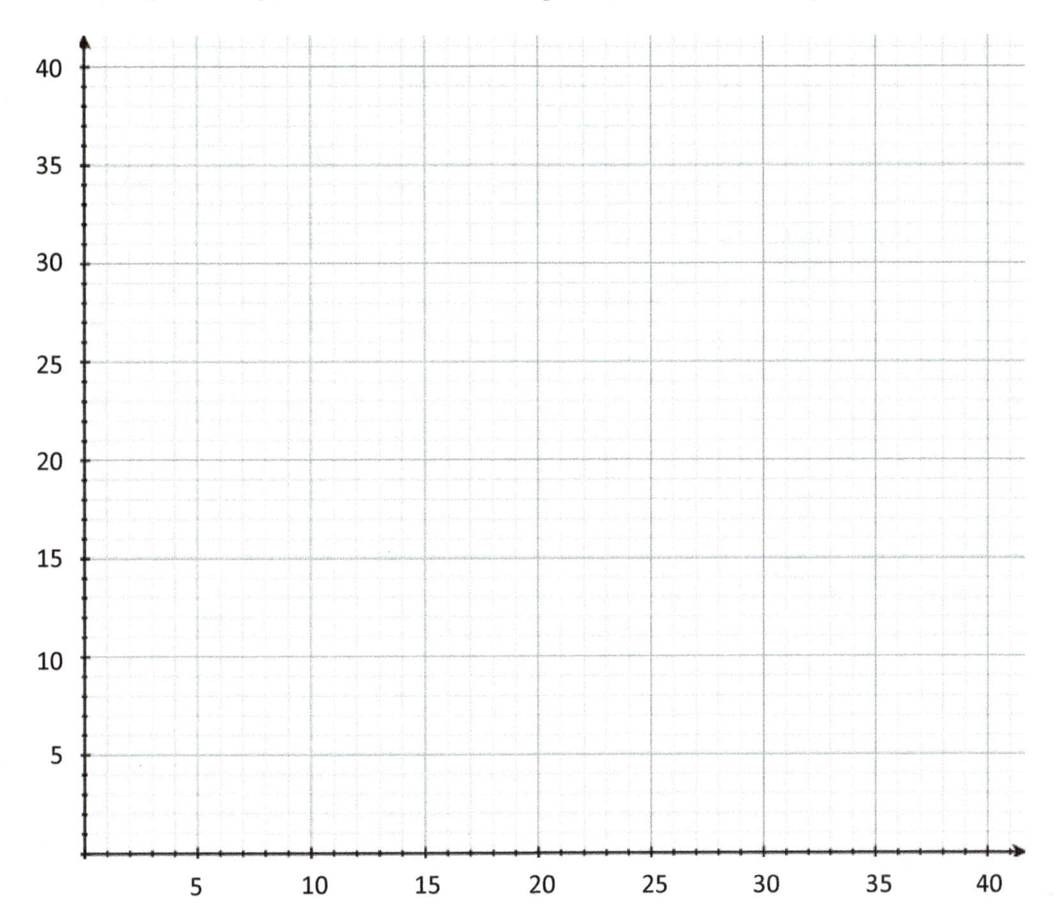

8. Use your graph to estimate the time it would take to travel 40 miles at 15 mi/hr.

9. Use your graph to estimate the speed needed to make the 40 mile trip in 3 hours.

10. (This one will impress your instructor.) Find an equation that relates speed to time for a 40 mile trip.

11. Use your equation to find how long it would take to make the trip at 50 miles per hour.

4-5 Class

In the Group portion of this lesson, we saw that for a 40 mile trip, as speed increases, the time of the trip decreases, and vice versa. This is typical of **inverse variation,** a relationship between two variable quantities that can be described by an equation of the form $y = \frac{k}{x}$, where k is some constant. Let's compare direct and inverse variation.

Comparison of Direct and Inverse Variation

The following are various ways of illustrating what it means to say that two variable quantities vary directly.

Verbally

The quantity y varies directly as the quantity x, and the constant of variation is k.

Algebraically

$y = kx$

Example:
$y = 3x$

Numerically

x	$y = 3x$
1	3
2	6
3	9
4	12
5	15

Graphically

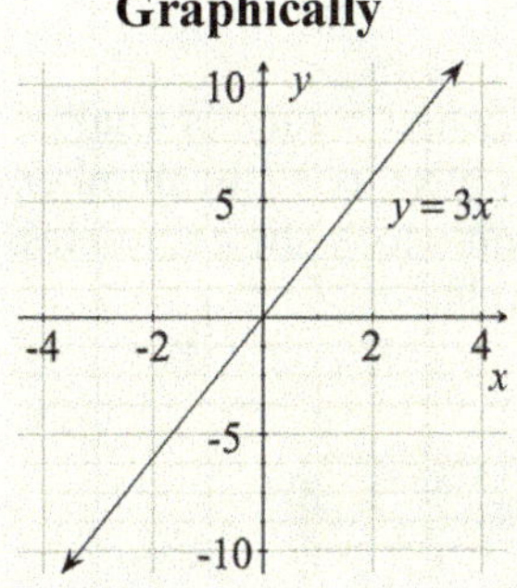

And these are various ways of illustrating what it means to say that two variable quantities vary inversely.

Verbally

The quantity y varies inversely as the quantity x, and the constant of variation is k.

Algebraically

$y = \frac{k}{x}$ for $x \neq 0$

Example:
$y = \frac{12}{x}$

Numerically

x	$y = \frac{12}{x}$
1	12
2	6
3	4
4	3
5	2.4

Graphically

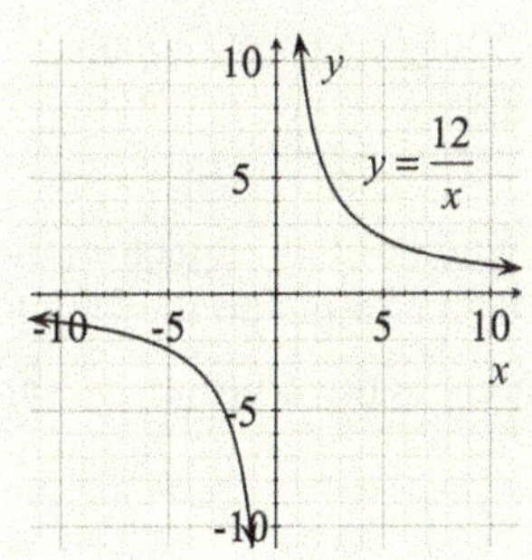

1. When you exchange currency, the value of currency you trade in varies directly with the amount you get back. So as the amount you trade in increases, the amount you get back ________________.

2. When you drive a certain distance, the time it takes varies inversely with your speed. So if your speed increases, the amount of time for the trip ________________.

In Questions 3-5, write an equation of variation for the situation, using k as the constant of variation.

3. An hourly worker's pay P varies directly as the amount of time they work T.

4. The pressure of propane in a tank P varies inversely as the volume of the gas V.

5. The illumination from a light source L varies inversely as the square of your distance d from the source.

6. A particular student's total points in a class P varies directly as the number of homework assignments h she completes. Use this to fill in the table.

h	P
8	48
16	
24	
32	
40	

7. The number of days d it takes for a construction crew to repave a certain road varies inversely as the number of workers w that work on the project. Use this to fill in the table.

w	d
8	48
16	
24	
32	
40	

8. The weight of an object on the Moon varies directly as its weight on Earth. If an astronaut weighs 150 lbs on Earth and 25 lbs on the Moon:

a. What is the constant of variation?

b. Write an equation relating the weight of an object on the Moon to its weight on Earth.

9. Use your equation from Question 8 to fill in the table of values.

Weight on Earth (lbs)	Weight on the Moon (lbs)
120	
135	
150	
165	
180	
195	

10. On one excursion to the Moon, the landing craft was twice as heavy when loaded on Earth as on the previous trip. How do the weights compare on the Moon?

11. On the next trip, the landing craft weighed 1/3 as much as the previous trip. How do the weights compare on the Moon?

12. When a wheel rolls along, the number of times it completes a full revolution to cover a certain distance varies inversely as the circumference of the wheel. Explain why this makes perfect sense.

13. The wheels on my Explorer have a circumference of just about 240 cm. If they make 100 revolutions in traveling a certain distance:

a. Find the constant of variation.

b. Write an equation relating the number of revolutions made by a wheel to its circumference in cm for this particular distance.

14. Use your equation from Question 13 to complete the table of values.

Circumference (cm)	Number of revolutions
100	
150	
200	
250	
300	

15. What happens to the number of revolutions if the circumference is tripled?

16. What happens to the number of revolutions if the circumference is cut in half?

4-5 Portfolio

Name ______________________________

Check each box when you've completed the task. Remember that your instructor will want you to turn in the portfolio pages you create.

Technology

1. ☐ Use Excel to create a table and graph illustrating the rate-time problem at the beginning of this lesson. You should use a formula to calculate the times corresponding to a variety of speeds. Begin by using a total distance of 40 miles. Bonus points if you create a second worksheet using a parameter that allows you to change the distance to anything you like in a single cell and have that automatically change the table and graph. There's a template to get you started in the online resources for this section.

Skills

1. ☐ Include any written work from the online skills assignment along with any notes or questions about this lesson's content.

Applications

1. ☐ Complete the applications problems.

Reflections

Type a short answer to each question.

1. ☐ Describe the differences between direct and inverse variation from as many different aspects as you can think of.
2. ☐ Explain why the topic of inverse variation is in this unit. (Hint: Look at the unit title.)
3. ☐ Take another look at your answer to Question 0 at the beginning of this lesson. Would you change your answer now that you've completed the lesson? How would you summarize the topic of this lesson now?
4. ☐ What questions do you have about this lesson?

Looking Ahead

1. ☐ Read the opening paragraph in Lesson 4-6 carefully, then answer Question 0 in preparation for that lesson.

4-5 Applications Name ____________________

For each problem, the game plan is to use the information to write an equation of variation. The information provided will allow you to find the constant of variation. Then use your equation to solve the problem.

1. The illumination provided by a car's headlight varies inversely as the square of the distance from the headlight. A headlight produces 15 foot-candles (fc) at a distance of 20 ft. What will the illumination be at 50 ft?

2. The amount of newsprint used by a newspaper varies directly as the number of people in the area served by that paper. According to the North American Newsprint Producers Association, the newsprint used to supply the annual needs of 1,000 people is 34,800 kg. How many kilograms would be needed to supply Columbus, Ohio, which has a population of about 720,000?

3. A child was struck by a car in a crosswalk. The driver of the car had slammed on his brakes and left visible skid marks on the pavement. He told the police he had been driving at 30 miles per hour. The police know that the length of skid marks L (when brakes are applied) varies directly as the square of the speed of the car v, and that at 40 miles per hour (under ideal conditions) skid marks would be 70 feet long. If the man is telling the truth, how long should the skid marks be?

4. The time it takes to get a sunburn varies inversely as the UV (ultraviolet light index) rating. According to CBS News online Consumer Tips, at a UV rating of 6, an average person can get a sunburn in as little as 15 minutes. How long would it take to get a sunburn when the UV rating is 10?

Lesson 4-6 Attraction and Melted Chocolate

Learning Objectives

- ☐ 1. Write large and small numbers in scientific notation.
- ☐ 2. Use scientific notation.

Mysteries of attraction could not always be explained by logic.
– Lisa Kleypas

Attraction: a mysterious topic that is like oxygen to the romance novelist and movie producer. We all feel emotional or physical attractions to certain people. But did you know that there's a real attraction between you and every person you come into contact with? It's called "gravity," and one of the physical laws of our universe is that any two objects have a gravitational attraction that can be quantified. This attraction is so tiny that it's hard to quantify (and you can't feel it) when the objects are small, like people. But when objects are really big, like the Earth and Sun, the gravitational attraction is very strong. Dealing with the force of gravity, by its very nature, requires working with both very small and very large numbers. Using regular decimal notation, this can be pretty cumbersome: do 0.0000004 and 0.000000004 look very different to you? One is a *hundred times as big* as the other. In this lesson, we'll use attraction and melted chocolate to study a special way of writing very small or very large numbers in a more readable, convenient way.

0. After reading the opening paragraph, what do you think the main topic of this section will be?

4-6 Class

Newton's law of universal gravitation describes the gravitational attraction between any two objects. The formula is

$$F = G\frac{m_1 m_2}{r^2}$$

where F is the force of the gravitational attraction, m_1 and m_2 are the masses of the two objects, and r is the distance between the centers of the two objects. G is a number known as the **gravitational constant:** we'll see what that number is in a bit.

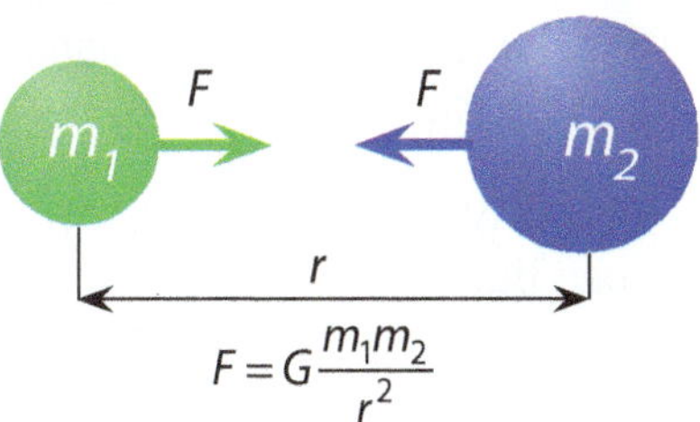

1. Newton's law of universal gravitation certainly looks like a variation equation, especially when you consider that G is a constant. Describe the relationship between the gravitational force between two objects and the distance separating their centers. Your answer should use either the term "directly" or "inversely," and should describe in plain English what happens to one of those quantities when the other changes.

2. Repeat Question 1, but this time describe the relationship between the gravitational force and the mass of the first object, m_1. (It might help to pretend that everything except F and m_1 are just numbers.)

3. Why can you feel the gravitational attraction between you and the Earth, but not between you and the person sitting next to you? (Notice that I said "gravitational attraction." 😉) Refer to your answer to Question 2.

4. Why can you feel the gravitational attraction between you and the Earth, but not between you and Jupiter? Refer to your answer to Question 1.

The force of gravity between the Earth and Sun is strong enough to keep us in orbit because even though we're about 149,597,870,700 meters apart, the mass of the Earth is 5,973,600,000,000,000,000,000,000 kilograms, and the mass of the sun is 1,989,000,000,000,000,000,000,000,000,000 kilograms. If you're like most people (including me, so don't feel bad) you have NO idea what those numbers actually mean. They're just too big to get any perspective on them other than "wow, those are really big." To make it easier to understand and work with numbers of ludicrous size, we'll use a clever method for writing them using exponents.

Consider the number 300, which can also be written as 3×10^2, which is of course 3×100. If you're thinking "that's silly," hang in there ... there is a point. Take a look at the first handful of powers of 10:

$$10^1 = 10 \qquad 10^2 = 100 \qquad 10^3 = 1{,}000 \qquad 10^4 = 10{,}000 \qquad 10^5 = 100{,}000$$

Do you see the pattern? Each extra power of 10 puts another zero on the decimal equivalent. In fact, the power indicates exactly the number of zeros included. Here's why this is helpful: Five billion is a tremendously large number: 5,000,000,000. Notice that there are nine zeros, so we can write five billion as 5×10^9, which is more concise and easier to read. If that makes sense to you, then you understand the idea behind *scientific notation*.

Writing Large Numbers in Scientific Notation

Example: 3,968,000

1. Write a decimal point at the end of the number, then move it left so that there is just one digit before it.

   3,968,000. (decimal moved six places)

   3.968 (the extra zeros are unnecessary)

2. Multiply the result by 10 to some power. The power will be the number of places you moved the decimal.

   3.968×10^6

5. Write each number in scientific notation.

a. The number of people in the United States in April 2013: 315,700,000

b. The distance in miles from Earth to the Moon: 384,400

We can also use scientific notation to write very small numbers, using negative powers of 10:

$$10^{-1} = \frac{1}{10} = 0.1 \qquad 10^{-2} = \frac{1}{10^2} = 0.01 \qquad 10^{-3} = \frac{1}{10^3} = 0.001 \qquad 10^{-4} = \frac{1}{10^4} = 0.0001$$

This time, the pattern is that when the negative exponent gets larger by one (meaning more negative), the decimal point slides one spot to the left. So, for example, 0.03 is the same as 3×10^{-2}.

Writing Small Numbers in Scientific Notation

Example: 0.000437

1. Move the decimal point right so that there is just one nonzero digit before it. Drop all zeros that come before that digit.

   0.000437 (decimal moved four places)

   4.37

2. Multiply the result by 10 to some power. The power will be the negative of the number of places you moved the decimal.

   4.37×10^{-4}

6. Write each number in scientific notation.

a. The weight in grams of one grain of spruce pollen: 0.00007

b. The average number of hours it took the U.S. Government to spend $1 million dollars in 2012: 0.0021605

In working with numbers in scientific notation, you should be able to convert them back to decimal notation.

Converting from Scientific to Decimal Notation

1. If the power of ten is positive, move the decimal point to the right the same number of places as the exponent. You might need to put in zeros as placeholders when you run out of digits. Putting in commas will make it easier to read a large number.

   Example: 4.01×10^8

   4.01 (decimal moved eight places)

   401,000,000 (fill in extra zeros)

2. If the power of ten is negative, move the decimal point to the left the same number of places as the exponent, again putting in zeros as needed.

   Example: 3.2×10^{-5}

   3.2 (decimal moved five places)

   0.000032

7. Write each number in decimal notation.

a. The number of drops in a gallon of water: 9.085×10^4

b. The mass of an average dust particle: 7.53×10^{-10} grams

Now let's see if you really understand scientific notation.

8. When a number written in scientific notation has a positive exponent on 10, what can we say for sure about the number?

9. If a number between 0 and 1 is written in scientific notation, what can we say about the exponent on 10?

10. Fill in the blanks: If a number in scientific notation has a negative exponent on 10, to convert to decimal form, move the decimal place ___________ because the number is __________________.

Using Technology: Scientific Notation

Spreadsheets and graphing calculators are both programmed to provide answers in scientific notation only if the size of the result is especially large or small.

TI-84 Plus Calculator

Notice that whether the numbers are entered in decimal or scientific notation, the result is in scientific notation only if it's very large. The calculator displays 1.2375×10^{11} as 1.2375E11.

To input 1.5×10^{12} in scientific notation:

Press 1.5 [2nd] [,] 12. This is listed as "EE" on the calculator.

If you want to force the results of calculations to be in scientific notation, put the calculator in scientific mode:

Press  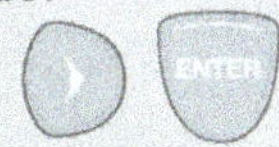 to set the calculator to scientific mode. The result:

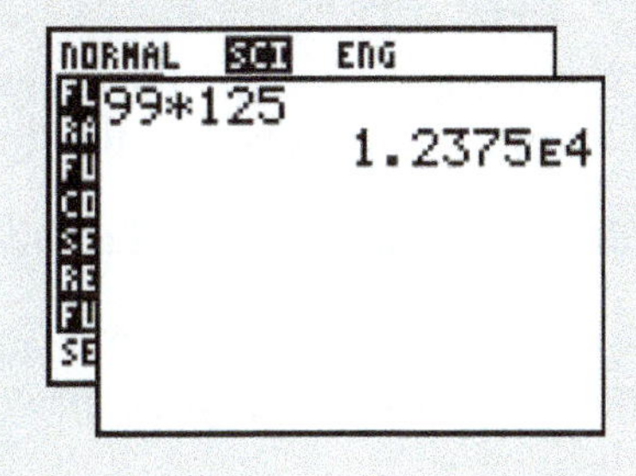

Excel

C1 f_x =A1*B1

	A	B	C
1	99	125	12375
2	990000	125000	1.24E+11
3	1.50E+12	3.20E-08	48000
4			
5	99	125	1.24E+04

Notice that in rows 1–3, Excel decides whether or not to put the result in scientific notation depending on its size.

To input 1.5×10^{12} in scientific notation, as in cell A3, type 1.5E 12. Excel will interpret this as scientific notation.

If you compare rows 1 and 5, you can see that the product is the same in each case: 99 * 125. The difference in the format of the output is based on the formatting option chosen for the cell. In cell C1, the **General** format was chosen from the **Number** menu, in which case Excel decides on the format as mentioned. In cell C5, **Scientific** was chosen from the **Number** menu, so the result is shown in scientific notation regardless of its size.

11. Write the result of the calculation on the calculator screen in scientific and decimal notation.

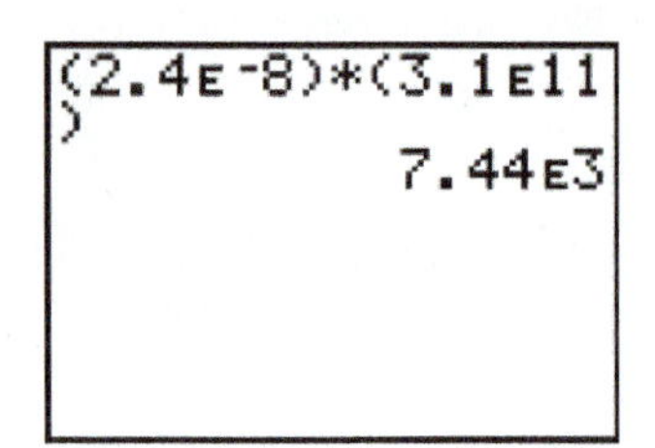

12. A typical MP3 file fills about 4×10^6 bytes of disk space. How many average songs can my iPod classic, which has 80 gigabytes of disk space, hold? Each gigabyte is 1×10^9 bytes.

13. Use Newton's law of universal gravitation to find the gravitational force between the Earth and Sun. The gravitational constant G is 6.674×10^{-11}; the distance from the Earth to the Sun is 1.496×10^{11} meters; the mass of Earth is 5.974×10^{24} kg; the mass of the Sun is 1.989×10^{30} kilograms. (The units of force are the kilogram meter per second squared, also known as the Newton.)

14. Newton's second law of motion is $F = ma$, where F is force, m is mass, and a is acceleration. For a situation where the force is caused by gravity, say when you drop a bowling ball off of a building, we can use the universal law of gravitation with $a = G\frac{m_1}{r^2}$. (Legal disclaimer: If you drop a bowling ball off of a building, you're a jerk.) Use the value of G, the mass of the Earth, and the radius of the Earth (6.3781×10^6 meters) to find the acceleration due to gravity near the Earth's surface. What are the units? How does your answer compare to the value used in a typical physics course, $a = 9.8$?

4-6 Group

A fun fact: you can use a microwave oven and a bar of chocolate to measure the speed of light. Really! Microwaves are electromagnetic waves, just like visible light. The difference is the wavelength of these waves: For visible light, the wavelengths are in the neighborhood of 10^{-6} m, while microwaves are close to 10^{-2} m. Basically, microwaves emitted in an oven make molecules in food vibrate, which causes heat. The first places to heat up are separated by half of a wavelength, so by using something that starts to melt visibly in certain spots, we can calculate the wavelength. We can then use that to calculate the speed of light, which I think we can all agree is pretty darn cool.

If your teacher is unusually adventurous, he or she might bring in a microwave and chocolate bars, in which case you can perform the experiment on your own. Start with around 20 seconds, and make sure you take out the carousel: the chocolate has to stay stationary. (And now you know why a lot of microwaves have a carousel: food heats more thoroughly at the spots corresponding to half of the wavelength.)

If no microwave and chocolate is available, I did the experiment at home and took a picture for you, which is reproduced here. It's a life-size scale, so you can measure the distance between the melted spots with a ruler.

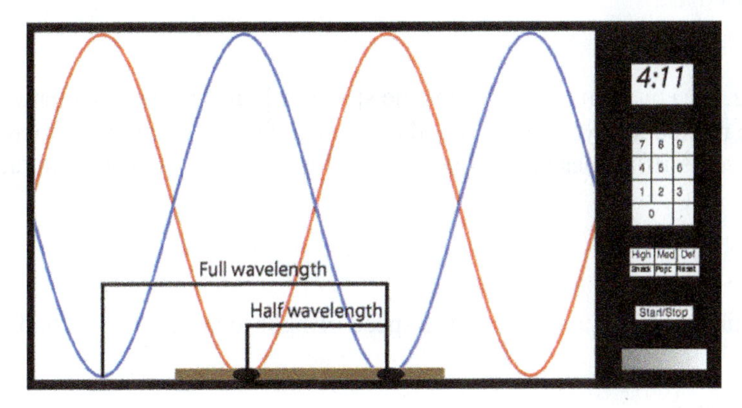

1. Measure the distance between melted spots on your chocolate, trying to measure from center to center. Record your distance below, remembering that this is half the wavelength of the microwaves.

Distance: _______________ cm

2. Convert the distance to meters. (Think about what the prefix "centi" means!)

Distance: _______________ m

3. Write the full wavelength: _______________ m

4. If you have any, eat the chocolate.

5. For electromagnetic waves, the equation $c = \lambda f$ relates the speed of light (c), the wavelength (λ), and the frequency (f), which is the number of cycles per second (Hz). The frequency of most microwaves is 2.45 GHz: one GHz represents 10^9 cycles per second. Write this number in scientific notation.

6. Use your answers to Questions 3 and 5 to calculate the speed of light in meters per second.

7. Use your phone or computer to look up the speed of light and write it here: _______________

8. Find the error and relative error in your estimate as a decimal, and as a percentage. (If you need a refresher, relative difference was introduced in Lesson 2-4.) If your answer is WAY off, you might want to check the microwave to see if the frequency is different from 2.45 GHz.

Math Note

There's a link to a really cool video about how microwave ovens work in the online resources for this section.

4-6 Portfolio

Name ______________________________

Check each box when you've completed the task. Remember that your instructor will want you to turn in the portfolio pages you create.

Technology

1. □ Watch the video on scale linked in the online resources for Lesson 4-6. Then tell everyone you know how cool it is.
2. □ After completing the Applications problems, create a spreadsheet that calculates the wavelength for the three FM radio stations you listen to most often. The spreadsheet should contain formulas in columns C and D to compute those values when you enter the information in columns A and B. A template to help you get started can be found in the online resources for this lesson.

	A	B	C	D
1	Radio Station	MHz	Hz	Wavelength
2	WLRW	94.5	94500000	3.172407

Skills

1. □ Include any written work from the online skills assignment along with any notes or questions about this lesson's content.

Applications

1. □ Complete the applications problems.

Reflections

Type a short answer to each question.

1. □ What is the point of scientific notation?
2. □ When a number is in scientific notation, how can you tell if it's really large or really small? Explain.
3. □ Take another look at your answer to Question 0 at the beginning of this lesson. Would you change your answer now that you've completed the lesson? How would you summarize the topic of this lesson now?
4. □ What questions do you have about this lesson?

Looking Ahead

1. □ Read the opening paragraph in Lesson 4-7 carefully, then answer Question 0 in preparation for that lesson.

4-6 Applications Name ______________________

Radio waves are also examples of electromagnetic waves. When you listen to a radio station, the frequency that you find on the dial describes the frequency of the waves sent out by that particular station. Since all electromagnetic waves move at the speed of light, you can calculate the wavelength for a station. For FM stations, like 94.5, this frequency is in Megahertz, which is millions of cycles per second. AM stations on the other hand, like 1100, are in Kilohertz, which is thousands of cycles per second.

1. Since c doesn't change in the formula $c = \lambda f$, what happens to f if λ increases? What happens to λ if f increases?

2. Describe how λ and f are related in terms of variation.

3. Write an equation that calculates λ if you know c and f. (Hint: Start with $c = \lambda f$.)

4. If you're listening to the Bob and Tom show on 94.7 FM, what is the frequency? Write it in scientific notation in terms of hertz.

5. What is the wavelength of the radio waves from this station?

6. How many times larger is your answer to Question 5 than the wavelength for microwaves calculated in the Group portion of this lesson?

7. Repeat Questions 4-6 for a baseball game on 1120 AM.

Lesson 4-7 Minding Your Business

Learning Objectives

- ☐ 1. Combine expressions using addition, subtraction, and multiplication.
- ☐ 2. Apply multiplication techniques to genetics.

An economist is an expert who will know tomorrow why the things he predicted yesterday didn't happen today. .
– Laurence J. Peter

At its core, business is simple: come up with a product or service that people want, and sell it to them. It gets complicated by competition, though. If you're the only person selling a product that people want or need, you're golden. But how often does that happen? Pretty much never. One of the most important aspects of business is studying financial figures and using them to predict future conditions. And since those figures are numbers, it should come as no surprise that we can use algebraic equations to model them. In this lesson, we'll use some common economic concepts (and an interesting topic in the field of genetics) to study some ways that we can combine expressions involving variables in order to model more complicated things in our world.

0. After reading the opening paragraph, what do you think the main topic of this section will be?

4-7 Group

From years of experience, the owner of a small boutique has found that she can sell $x = 300$ paris of earrings per month when she charges $p = \$15$ per pair. She's able to sell 375 pairs per month at a price of \$12 per pair.

1. If the relationship between price and sales is linear, write an equation that relates the price p to the number of pairs of earrings sold x. Your equation should be in the form $p = mx + b$. Hint: Start by writing ordered pairs of the form (earrings sold, price), then find the slope.

In economics, the word **revenue** refers to the amount of money that a business takes in as a result of selling a product or service. Revenue is pretty easy to calculate, especially if you know the amount of sales and the price: revenue is simply the selling price times the number of units sold. Using the symbols from Question 1, revenue R is number of units sold x times price p.

2. Write an equation that represents the revenue made on earrings for our boutique owner by using the equation $R = xp$. Then simplify your expression by multiplying out the parentheses using the distributive property.

The **costs** of doing business for a company can be found by adding fixed costs, like rent, utilities, insurance, and wages, and variable costs, which is what it costs to produce the items being sold. These costs are variable because the total varies depending on the number of items produced.

3. The portion of the boutique's fixed monthly costs attributed to earrings is $530; the supplies needed to produce the earrings cost $2.20 per pair. Write an equation that represents the total cost C to the boutique of producing x pairs of earrings. Your equation should be in the form $C = mx + b$.

4. Since revenue is the amount of money a business takes in, and cost is the amount they pay out, the difference between revenue and cost is the **profit** made by the business. That is, profit P is revenue R minus costs C. Write an equation that represents the profit made on earrings by the boutique. (Hint: Substitute your answers to Questions 2 and 3 into the formula $P = R - C$.) Don't simplify your expression yet.

5. The costs consist of two distinct terms, and when we subtract we need to make sure to subtract ALL of the costs, not just some of them. To make sure this happens, we put parentheses around the entire cost expression; if you didn't already do this, add parentheses to your answer from Question 4. Then perform the subtraction, keeping in mind that the parentheses are there to remind you to subtract EVERY term in the expression for cost. (A look back at what we learned about "like quantities" in Lesson 1-5 might prove useful.) The result is an equation describing profit in terms of pairs of earrings sold that's relatively easy to work with.

Math Note

In algebra, **terms** are individual parts of an expression that are added or subtracted. For example, in $250x + 30$, the terms are $250x$ and 30.

6. Find the profit the boutique makes if it sells 200 pairs of earrings per month.

7. Find the profit made from selling 500 pairs per month.

8. What's the profit if the boutique sells no earrings in a month? What does the result mean?

9. Use trial and error to find the number of pairs of earrings sold per month that will give the boutique its largest possible profit on earrings. This would make the owner happy.

10. What's the largest monthly profit that can be made on earrings? (Use your answer to Question 9.)

11. What selling price results in the largest possible profit?

12. With profits soaring due to your economic expertise, the boutique owner decides to open a second location, and finds that the profit equation for selling earrings at the second location is $P = -0.05x^2 + 22x - 450$. Write a single equation that describes the combined profit on earrings for both locations.

The traits that parents pass on to their offspring are determined by genes, and the study of how genes affect individual traits is called genetics. Many traits are determined by a pair of genes in an individual, with one gene coming from each parent. An example is the disease cystic fibrosis. Let's call the disease-causing variation of the gene f, and the one that doesn't cause the disease F. These gene variations are called **alleles.** If you have one copy of each, you won't have the disease: the allele for not having it is called **dominant** because just one copy of it will make the trait (not having cystic fibrosis) appear. The disease-causing allele is called **recessive,** and the disease manifests only if the offspring gets two copies of that allele. So any of Ff, fF, or FF will result in a child without cystic fibrosis, but if the child receives ff, he or she will have the disease.

The result is that you can pass on cystic fibrosis to a child even if you don't have the disease, as long as you have one copy of the recessive gene f. In that case, you're called a **carrier** of the recessive trait. We used capital F for the dominant gene as a reminder that it's dominant, and if one copy appears its trait appears. To show the likelihood of a trait being passed on, we can use a diagram known as a **Punnett square.** The alleles belonging to each parent are written across the top and down the left side: the results of those alleles being passed on are shown in the interior boxes.

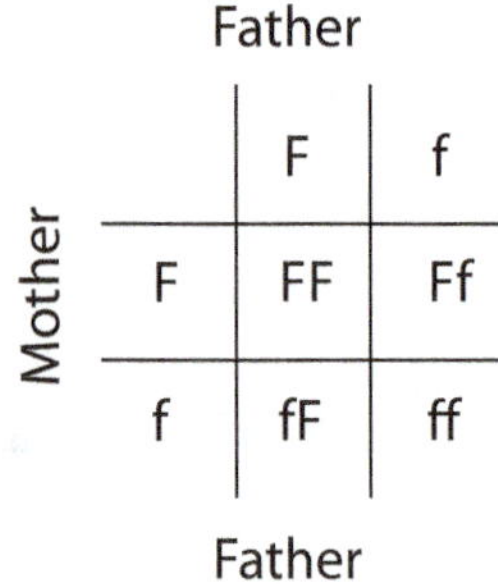

In this case, the parents are both carriers: their genes are both Ff. They can each pass on the F allele, in which case the child will be healthy and not a carrier. If one parent passes on the F allele and the other the f allele, the child will be a carrier. And if both parents pass on the f allele, the child will have cystic fibrosis. In fact, assuming that each parent is equally likely to pass on either F or f, this Punnett square shows that there's a 25% chance that a child conceived by these parents will have cystic fibrosis, and a 50% chance that the child will be a carrier.

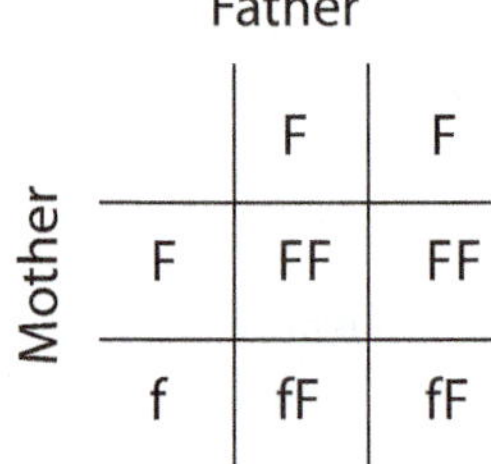

Now one of the parents is not a carrier, which changes the picture considerably. Two of the four squares are fF, meaning that the child will be a carrier. The other two are FF, meaning the child will be clear of cystic fibrosis altogether. Most importantly, these parents would have a 0% chance of having a child with cystic fibrosis.

13. Fill in the two Punnett squares below for some other genetic combinations for two parents, and discuss what the results say about their children.

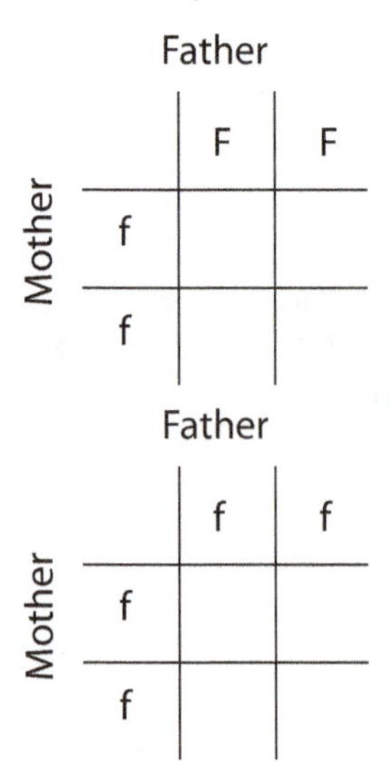

You might be wondering what genetics has to do with the topic of this section. It turns out that an idea similar to the Punnett square can be used to develop a method for multiplying expressions that consist of two terms, like $x + 4$.

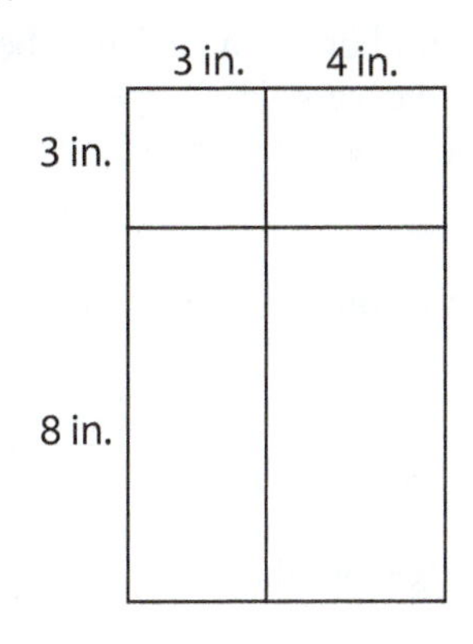

14. Find the area of the large rectangle by first multiplying the overall length and the overall width.

15. Now find the area of the large rectangle by finding the area of each of the 4 smaller rectangles and adding them together.

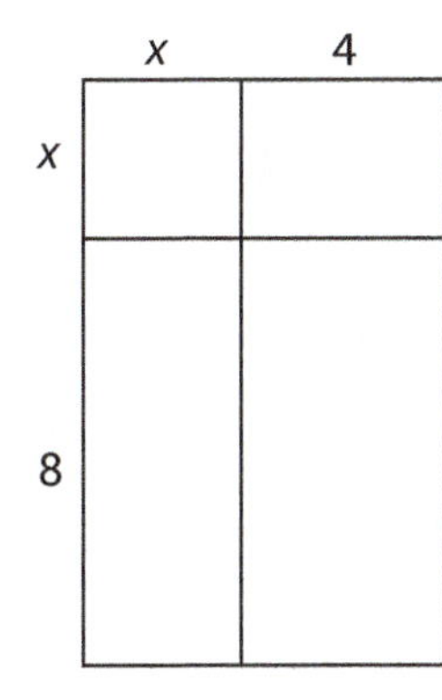

16. Find the area of the large rectangle by multiplying the overall length and the overall width. Your answer will be an algebraic expression: leave it in product form.

17. Now find the area by finding the area of each of the 4 smaller rectangles and adding them together.

18. Use the technique of Questions 16 and 17 to find the result of multiplying $(x + 3)(x + 5)$.

The form $(x + 3)(x + 5)$ is called the **factored form** of an expression. Notice that it's written as a product: $(x + 3)$ TIMES $(x + 5)$. Your answer to Question 18 is called the **expanded form** of an expression, and it's written as a sum or difference. These are two different forms of the same expression: if you input ANY value for the variable quantity x, you should get the same result from either expression. (We say that the expressions are **equivalent.**)

19. How could you use a graphing calculator or spreadsheet to verify that the two forms of $(x + 3)(x + 5)$ are really equivalent? Explain, and perform the check you described.

4-7 Class

An algebraic expression consisting of one or more terms that look like either a real number, or a real number times a variable raised to a whole number power is called a **polynomial.** In this lesson, we've studied operations on polynomials: combining polynomials using addition, subtraction, and multiplication. Let's summarize what we've learned.

In Questions 1-4, simplify each expression. Aside from what we covered in the Group portion of this lesson, here are some things you should keep in mind:

- Order of operations: multiplication always comes before addition or subtraction.
- The distributive property is needed when multiplying a number or variable by a sum or difference.
- When subtracting, don't forget to subtract EVERY term in an expression. That's what the parentheses are there for!

1. $\left(3x^2+3x+5\right)+\left(2x^2-3x-10\right)$

2. $\left(5x^2+3x+2\right)-\left(x^2-2x+7\right)$

3. $3\left(x^2+3x+5\right)+2\left(x^2-3x-10\right)$

4. $2\left(3x^2+4x-5\right)-3\left(4x^2-5x+1\right)$

Multiplying two terms with variables is very similar to multiplying dimensions with units.

5. Perform each multiplication. For parts a and b, include units.

a. 2 ft × 3 ft = _______

b. 6 ft² × 5 ft = ________

c. $(2x)(3x)$ = _______

d. $(6x^2)(5x)$ = _______

The multiplication that we did by borrowing the concept of Punnett squares was an example of multiplying polynomials. In much the same way that each of the alleles for each parent was combined with each of the alleles from the other, when multiplying polynomials with two terms, each of the terms in the first polynomial was multiplied by each of the terms in the second, then all of the resulting products were added. Does that also work for situations other than two terms times two terms? Before tackling that question, let's get more practice at the type of multiplication describe in this paragraph.

6. Perform each multiplication:

a. $(x - 5)(x + 7)$

b. $(2y + 6)(3y - 8)$

7. Use the distributive property to multiply: $2x(3x + 4)$. Does this match the description in the above paragraph? Explain.

8. The goal is to multiply $(x - 5)$ by $(x^2 + 3x + 4)$.

a. First, multiply x by each term of $x^2 + 3x + 4$, and list the three products.

b. Now multiply -5 by each term of $x^2 + 3x + 4$, and again list the three products. (Careful with the negative sign!)

c. Add all six products from parts a and b, combining any like terms. This is (we hope!) the product of $(x - 5)$ and $x^2 + 3x + 4$.

9. Using a graphing calculator or computer graphing program, graph both $(x - 5)(x^2 + 3x + 4)$ and your answer from Question 8c. Do you think the two expressions are equal? Explain.

Summary:

To multiply one polynomial by another, multiply each __________ of the first polynomial by each __________ of the second polynomial, then combine any __________ __________.

4-7 Portfolio

Name ______________________________

Check each box when you've completed the task. Remember that your instructor will want you to turn in the portfolio pages you create.

Technology

1. ☐ Create an Excel spreadsheet to explore whether or not your answer to Question 8c in the Class portion of this lesson is correct. Make sure to input a wide variety of x values: positive, negative, fractions, decimals, even irrational numbers if you want to impress your teacher. Can you be sure that the two expressions are equivalent by inputting a variety of values? Explain. (Bonus points for doing some online research about when two polynomials are equivalent ... .) There's a template to help you get started in the online resources for this lesson.

	A	B	C
1	Input (x)	(x-5)(x^2+3x+4)	Answer to 8c
2			
3			
4			

Skills

1. ☐ Include any written work from the online skills assignment along with any notes or questions about this lesson's content.

Applications

1. ☐ Complete the applications problems.

Reflections

Type a short answer to each question.

1. ☐ What is a polynomial? What is meant by the phrase "operations on polynomials"?
2. ☐ Describe the significance of each of these words in the context of this lesson: revenue, costs, profit.
3. ☐ Take another look at your answer to Question 0 at the beginning of this lesson. Would you change your answer now that you've completed the lesson? How would you summarize the topic of this lesson now?
4. ☐ What questions do you have about this lesson?

Looking Ahead

1. ☐ Read the opening paragraph in Lesson 4-8 carefully, then answer Question 0 in preparation for that lesson.

4-7 Applications Name ____________________

Based on the results of an extensive market research study that I totally made up, an economist hired by a company that makes a popular line of inexpensive coffee makers predicts that consumers in one market will buy x units per week of one model if the price is $p = -0.27x + 51$ dollars.

1. Recall that revenue is price times number of units sold, and write an equation that represents the revenue the company will make from selling x units of the coffee maker in that market. Simplify your answer.

2. It costs the manufacturer $C = -0.12x^2 + 12x + 720$ dollars to make x units of this particular coffee maker. Write an equation describing the profit made from selling x units of the coffee maker in this market. The simpler your answer is, the easier it will be to work with in the next couple of questions.

3. Find the profit made in the market if 20 coffee makers are sold per week. What does this mean?

4. Find the profit made for sales of 100 and 200 coffee makers. How do they compare?

5. Use trial and error to estimate the number of units sold that will lead to the highest profit in this market, then find the price that will accomplish this result.

4-7 Applications Name ____________________

Evidence suggests that immunity to poison ivy might be a dominant trait, even though it's relatively unusual. Let's assume that this is true, and use P to represent the dominant allele (immune to poison ivy) and p to represent the recessive allele (susceptible to poison ivy).

Here's a Punnett square for two parents who each have one dominant and one recessive allele in this situation. Assuming that each parent is equally likely to pass on either allele...

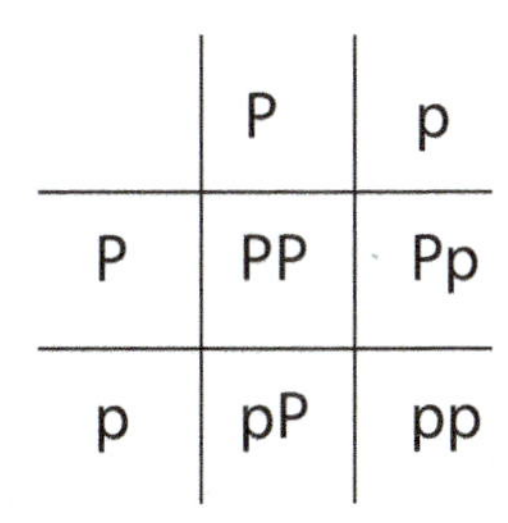

	P	p
P	PP	Pp
p	pP	pp

6. What's the percent chance that a child born to these parents is immune to poison ivy?

7. What's the percent chance that the child is a carrier for the recessive trait, but is immune?

8. What does the square pp represent in terms of poison ivy exposure?

9. Draw a Punnett square for one parent who is susceptible to poison ivy and one that is a carrier for that recessive trait.

10. What is the percent chance that a child born to the two parents in Question 9 will be susceptible to poison ivy?

Lesson 4-8 The F Word

Learning Objectives

- ☐ 1. Understand what factoring is and why it's useful in algebra.
- ☐ 2. Use function notation.
- ☐ 3. Study the connection between zeros and *x* intercepts.
- ☐ 4. Factor expressions.

Confidence is the most important single factor in this game, and no matter how great your natural talent, there is only one way to obtain and sustain it: work.

– Jack Nicklaus

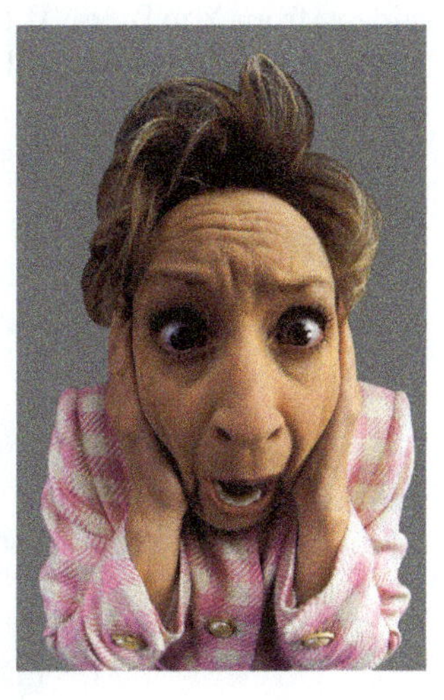

Relax – the F word we're talking about is perfectly acceptable by FCC standards. It's only in math classes that some people feel it's a dirty word: factoring. Depending on who you ask, factoring can be thought of as anything from an essential part of every college student's curriculum, without which they would surely perish, to an archaic procedure with very limited and artificial applications in the modern world. The truth is that you can type any polynomial into Wolfram Alpha and get its factored form. The trick is in understanding what the point is, and being able to efficiently use the factored form of a polynomial to solve problems. So that will be our first order of business. We'll look at the actual process of factoring later in the lesson, just in case your school is in the "you really, really, really, really need to be able to do this" camp.

0. After reading the opening paragraph, what do you think the main topic of this section will be?

4-8 Group

1. Perform the multiplication: $5 \times 7 =$ ________

2. Turn your answer from Question 1 around backwards to write 35 as a product: $35 =$ ______ $\times$ ______

Congratulations – you have just factored. Isn't it bizarre that a topic that has such a bad reputation is based on such a simple idea?

Definition of Factoring

Factoring is simply the process of writing a number or expression as a product. In that regard, it's essentially the opposite of performing multiplication.

Here's an example from our study of economics. In Lesson 4-7, we multiplied a variable quantity (number of units sold, represented by x) by the expression $-0.27x + 51$, which represented price, to get an equation describing revenue: $R = -0.27x^2 + 51x$. If we "undo" that multiplication to recover the price:

$$-0.27x^2 + 51x = x(-0.27x + 51)$$

we have factored, because we've written $-0.27x^2 + 51x$ as a product. Simple!

As you probably know, computers have a master volume setting that controls the volume of everything it sends through the speakers. But many applications have volume control as well, like web browsers when playing videos from YouTube. If the master volume is set to 100% and the volume in the YouTube window is set to 60%, the sound in the video will play at 60% of its max. Easy. Now let's look at some other combinations.

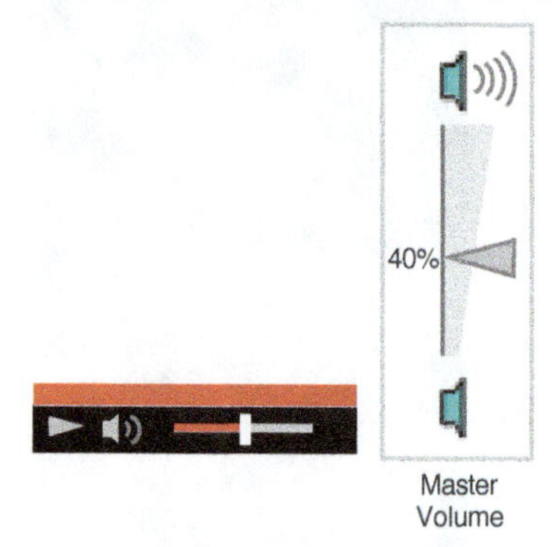

3. If the volume in the YouTube window is set at half and the master volume for the computer is at 40%, what percent of the max volume do you think you'll hear from your speakers? Explain your answer.

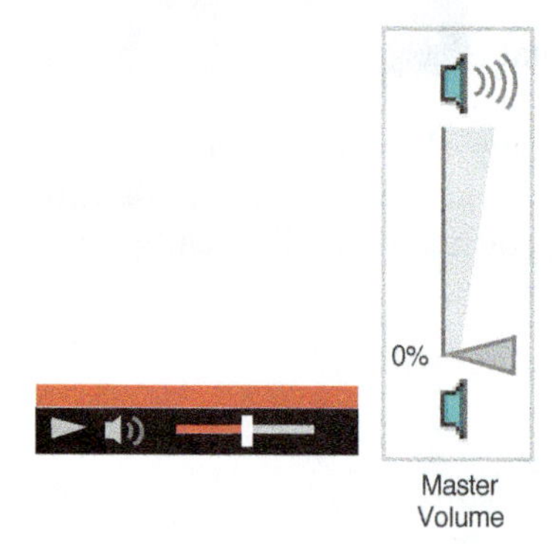

4. If the volume in the YouTube window is set at half and the master volume for the computer is at 0, what percent of the max volume do you think you'll hear from your speakers? Explain your answer.

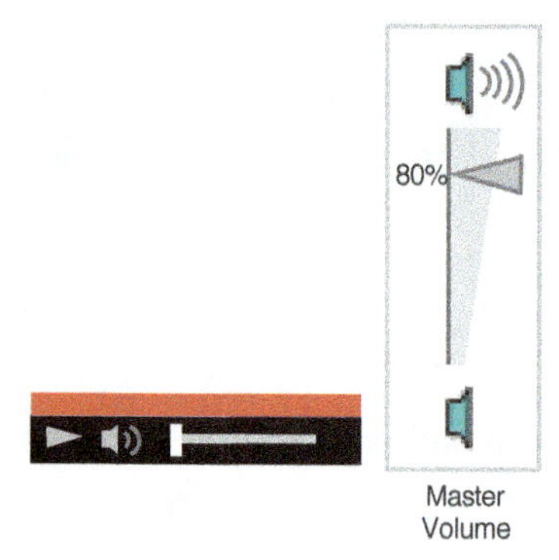

5. If the volume in the YouTube window is set all the way to the left and the master volume for the computer is at 80%, what percent of the max volume do you think you'll hear from your speakers? Explain your answer.

Conclusions:

6. To get the overall volume, we ____________________ the percentages from the master volume and the YouTube volume.

7. You can only get zero as the overall volume if ____.

So what does this have to do with factoring?

8. For what value of the variable quantity x does the expression $x - 4$ have output zero? What about the expression $x + 5$? (Don't over-think these. They're really easy.)

9. Now look at the product $(x - 4)(x + 5)$, which is a polynomial in factored form. Find the value of the product when...

a. $x = 2$ **b.** $x = 4$ **c.** $x = 0$ **d.** $x = -1$ **e.** $x = -5$

Conclusion:

10. The only way a product of algebraic expressions can have value zero is if ________________________

____.

And THAT'S why factoring is a big deal in algebra. It provides an important technique of solving equations that have zero on one side.

11. Write the number 37 as a product.

12. Write the polynomial expression $3x + 5$ as a product.

In Questions 11 and 12, there's only one way to write the given number or expression as a product (unless you're willing to use fractions, which we are not at the moment): one times the product or expression itself. In the case of 37, we call it a **prime number.** For the polynomial $3x + 5$, we say that it is a **prime polynomial.**

The point: Not every expression can be factored if we restrict our attention to integer coefficients.

4-8 Class

Without realizing it, we've been working with an important idea in math for quite some time now. Finally, we'll give it a name and symbols.

Definition of a Function

A **function** is a relationship between two quantities where each input produces a unique output.

If you make \$20 per night selling hot dogs at a ball game, plus 10% of your sales, your total pay can be modeled by the equation $y = 0.10x + 20$. We can think of this as a function: for each input x (amount of sales), there will be a unique output y (the amount of money you make). In order to make it clear that a given equation actually defines a function, we use **function notation.**

Function Notation

The output of a function named f is described by the symbol $f(x)$, which is read "f OF x", NEVER "f TIMES x."

Important note: $f(x)$ does NOT indicate multiplication; it's a way to indicate that the function f determines the output for a given input x.

For example, the function describing your pay from selling hot dogs could be written as $f(x) = 0.10x + 20$. In that case, $f(50)$ represents the output of the function (amount of money you make) when the input is 50 (meaning 50 dollars in hot dog sales). Let's see how you'd do:

$$f(50) = 0.10(50) + 20 = 5 + 20 = 25$$

Notice that in its original notation, the function is represented by $f(x)$, and then a formula is described that has the variable quantity x in it. In the notation $f(50)$, the symbol x has been replaced by the number 50. This indicates that we should do the same in the formula: replace the symbol x with 50. We now know that $f(50) = 25$, which means that if you sell \$50 worth of hot dogs, you'll make \$25.

1. For the hot dog sales function, find each function value (output) and describe what it tells you.

a. $f(0)$

b. $f(40)$

c. $f(100)$

d. $f(250)$

If $x = c$ is an input value for which the output of a function $f(c)$ is zero, we call the input c a **zero** of the function. There's a useful relationship between the factors of a polynomial, the zeros of the related function, and the x intercepts of the graph.

The Connection Between Factors, Zeros, and Intercepts

For a given polynomial $P(x)$ and a real number c, each of these statements say the same thing:

The point $(c, 0)$ is an x intercept of the graph of $P(x)$.
(Graphical)

The input c is a zero of $P(x)$,which means that $P(c) = 0$.
(Numerical)

The expression $(x - c)$ is one of the factors of $P(x)$.
(Algebraic)

2. For the polynomial $P(x) = (x - 1)(x + 5)$, find each output. Use P in this factored form.

a. $P(-5)$

b. $P(-3)$

c. $P(-1)$

d. $P(1)$

3. Use the table and graph provided for $P(x) = (x - 1)(x + 5)$ to answer each question.

x	P(x)
−5	0
−4	−5
−3	−8
−2	−9
−1	−8
0	−5
1	0

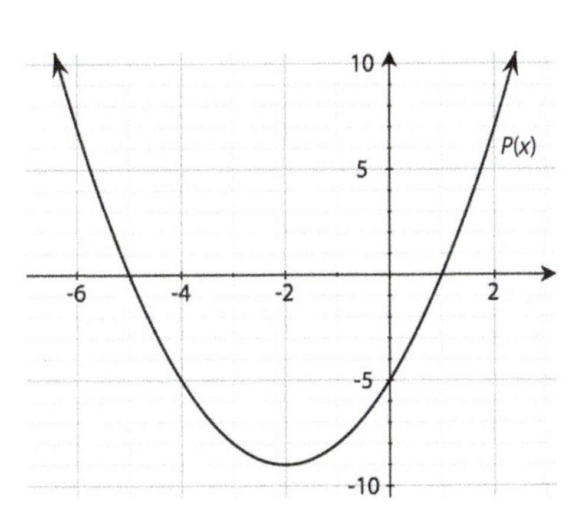

a. List the factors of $P(x)$.

b. List the zeros of $P(x)$.

c. List the x intercepts of the graph.

4-8 Class

4. Now let's look at the function $f(x) = x^2 + 2x - 3$.

x	$f(x)$
−4	5
−3	0
−2	−3
−1	−4
0	−3
1	0
2	5

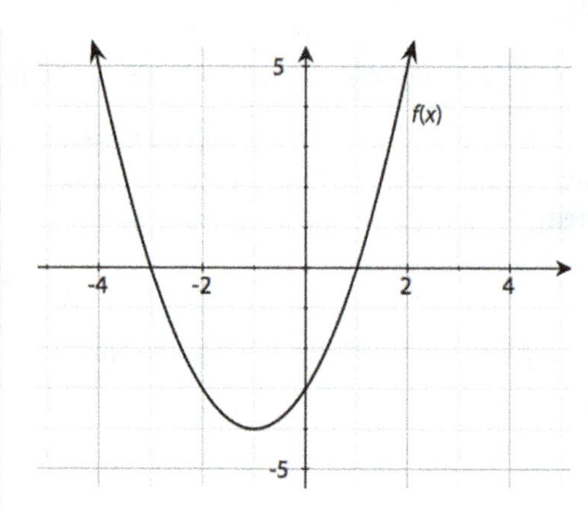

a. List the zeros of $f(x)$.

b. List the x intercepts of the graph.

c. Use parts a and b to write the factored form of $f(x) = x^2 + 2x - 3$.

Use the factored form of each polynomial to list the zeros and x intercepts of the polynomial's graph.

5. $P(x) = 2(x-3)(x-4)$

Zeros:

x intercepts:

6. $y(x) = (x+5)(x+4)(x-2)$

Zeros:

x intercepts:

Punnett Square Factoring

In the last lesson, we saw how the idea behind Punnett squares can help us to multiply polynomials. Since factoring is the process of undoing multiplication, it seems reasonable that we should be able to use Punnett squares to help with that too. And you know what? We can.

7. Remember that multiplying polynomials amounts to multiplying each term in the first polynomial by each term in the second, then combining like terms. Multiply $(x + 2)(x + 12)$ using the Punnett square to list each of the four terms.

	x	12
x		
2		

If we try to think about doing this process backwards, we might come to a realization: the first term of the product comes from the upper left box, and the last term comes from the lower right box. The combination of the other two boxes provides the middle term. So if we want to undo the process of multiplication we could set up a Punnett square like this:

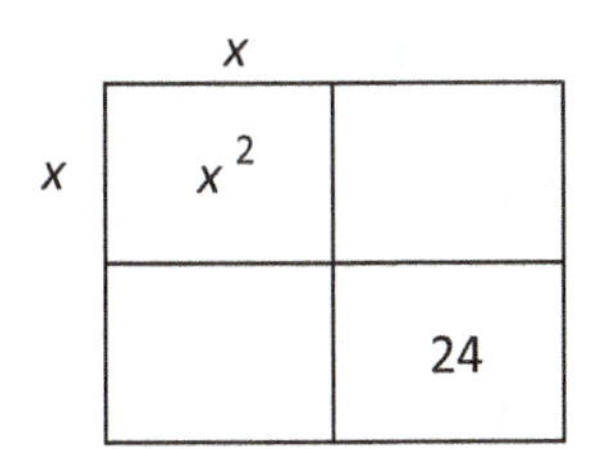

If we can fill in the blanks on the outside of the boxes, we can find the factored form. Our first job would be to find two numbers whose product is 24.

8. Write all pairs of whole numbers whose product is 24. (Remember, "product" refers to multiplication.)

9. Find a pair of numbers on your list from Question 8 that you can add or subtract to get 14. Is there more than one pair?

10. What is the connection between that pair of numbers and the factored form $(x + 2)(x + 12)$? What can you conclude?

11. Use the Punnett square to find the factored form of $x^2 + 7x + 10$.

	x	
x	x^2	
		10

12. Enter $x^2 + 7x + 10$ next to Y_1 on the **Y** = screen of your graphing calculator. Then press **ZOOM** 6. Where does the graph appear to cross the x axis? What does this tell you about the factors of $x^2 + 7x + 10$? Does this match the factored form you found in Question 11?

13. How can you check your answer to Question 11 using multiplication? Do it.

Next, we'll factor $y^2 + 7y - 18$ using a Punnett square.

	y	
y	y^2	
		−18

14. Write all pairs of integers whose product is −18.

15. Do any of those pairs combine to give you +7?

16. Use the result of Question 15 to fill in the square and factor $y^2 + 7y - 18$.

Math Note

When some coefficients are negative, make sure you include the negatives in your Punnett square.

What about if the coefficient of the squared term isn't one? Let's check that out next by factoring $2x^2 - 3x - 20$.

	x	-1
$2x$	$2x^2$	
20		-20

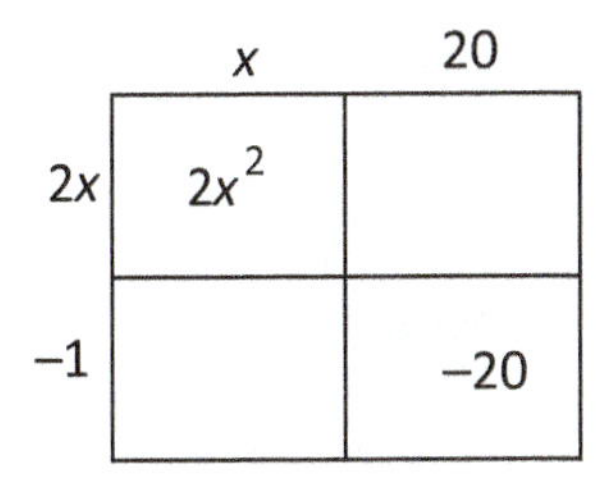

17. I've already filled in some possible pairs of numbers that will give us the -20 we need in the bottom right box. Complete each box and see if either of these combinations provide the correct factored form.

18. Draw Punnett squares that have all other possible pairs of numbers whose product is the -20 we need. Then find one that correctly provides the factored form for $2x^2 - 3x - 20$, and write the result here.

Dude, what's the point?

I'm glad you asked. We have seen that when you know the factored form of a polynomial, it's super easy to find the x intercepts of its graph. And this can be used to solve many problems, like this one.

If you make your living from growing produce, say apples, the more apples you grow, the more you can sell. Duh. So why not just plant a million apple trees? Space and resources. If the trees are too densely planted, they compete with each other for resources, and each tree produces less. So too many trees could mean less apples. And it's pretty obvious that not enough trees means less apples. Hey–maybe it's useful to find the ideal number of apple trees to plant.

In one orchard, there are 50 trees, each of which produces 800 apples. For every additional tree planted, the average output per tree drops by 10 apples. Bummer. How many additional trees should be planted to get the largest total number of apples?

19. If x more trees are planted, how many trees will there be total?

20. If the current yield per tree is 800, and will go down by 10 for each new tree planted, what will the yield per tree be if 5 more trees are planted?

21. If the current yield per tree is 800, and will go down by 10 for each new tree planted, what will the yield per tree be if x more trees are planted?

22. The total number of apples produced will be the total number of trees times the yield per tree. Write a function $T(x)$ describing the total number of apples, using your answers to Questions 19 and 21.

23. Find the x intercepts of the function you wrote in Question 22.

24. The graph of this function is a parabola that opens down, and parabolas are symmetric. That means the high point we're looking for has to be halfway between the x intercepts. How many new trees should they plant, and how many apples will be produced?

4-8 Portfolio

Name ______________________________

Check each box when you've completed the task. Remember that your instructor will want you to turn in the portfolio pages you create.

Technology

1. ☐ As we mentioned in the opening paragraph of this lesson, the website Wolfram Alpha (the URL, shockingly, is wolframalpha.com) can factor pretty much anything that is factorable. So go to the site and ask Wolfram Alpha to factor the polynomial from Question 18 in the Class portion of this lesson. In exchange for your effort, you'll get back a factored form of the polynomial, and a whole lot more. Put a printout of the result in your portfolio, then type a brief report on some of the other things you can learn about the polynomial this way.

Skills

1. ☐ Include any written work from the online skills assignment along with any notes or questions about this lesson's content.

Applications

1. ☐ Complete the applications problems.

Reflections

Type a short answer to each question.

1. ☐ What is the value of having the factored form of a polynomial? Think of as many different aspects as you can.
2. ☐ Describe the relationship between factoring and multiplication.
3. ☐ Take another look at your answer to Question 0 at the beginning of this lesson. Would you change your answer now that you've completed the lesson? How would you summarize the topic of this lesson now?
4. ☐ What questions do you have about this lesson?

Looking Ahead

1. ☐ Read the opening paragraph in Lesson 4-9 carefully, then answer Question 0 in preparation for that lesson.

4-8 Applications Name ______________________________

A trendy nightclub has a \$10 cover charge, and gets 660 patrons on an average night. After surveying customers, they're able to estimate that for each additional \$1 they add to the cover charge, they'll lose an average of 30 patrons per night.

1. If they add x one-dollar increments to the cover charge, write an expression describing the new charge.

2. If they add x one-dollar increments, how many patrons will they lose on average? What will be the new average number of customers?

3. Write a function $R(x)$ that describes the average nightly revenue if they raise the current price by \$$x$. (Hint: You'll need your answers to Questions 1 and 2.)

4. Find $R(0)$ and describe what it means.

5. Find the x intercepts of your function and describe what each tells us about the nightclub's cover charge and revenue.

6. If all has gone well, the graph of your revenue function should be a parabola that opens down. Since parabolas are symmetric, the highest point on the graph is halfway between the x intercepts. Use that fact to find the highest point on the graph, and describe in detail what it tells you about the nightclub's cover charge and revenue.

Lesson 4-9 Going... Going... GONE!

Learning Objectives

- ☐ 1. Solve equations using the quadratic formula.
- ☐ 2. Find the vertex of a parabola.
- ☐ 3. Study physical phenomena using quadratic functions.

Every strike brings me closer to the next home run.

– George Herman "Babe" Ruth

Baseball purists will tell you that a well-pitched game is the pinnacle of the sport, but the average fan goes to the park hoping to see some home runs–and the longer, the better. There's just something cool about seeing a ball hit back back back, and gone. And in the information age, fans don't just want to speculate on how far a home run was hit: They want a measurement, and want it RIGHT NOW. So most ballparks and TV broadcasts now supply a distance after a home run is hit. But how do they calculate the distances? If you guessed "math," you're pretty smart. We've already seen that objects flying through the air tend to follow a parabolic path. In this lesson, we'll learn more about studying parabolas from an algebraic standpoint. This allows us to get more information about things like–you guessed it–how far a home run travels.

0. After reading the opening paragraph, what do you think the main topic of this section will be?

4-9 Class

As of this writing it's been almost 5 years since a major league baseball player hit a home run longer than 500 feet in a game. That feat was accomplished by Arizona Diamondback Adam Dunn, and based on information provided by a cool website called "Home Run Tracker," we can approximate the flight of that ball with this function: $f(x) = -0.00132x^2 + 0.659x + 4.1$. The input x represents the number of horizontal feet from where the ball was hit, while the output $f(x)$ is the height above the ground after the ball had traveled x feet horizontally. In order to find the total distance the ball was hit, we'd need to find the horizontal distance (x) that corresponds to a height of zero feet (when the ball would have returned to ground level if it hadn't been stopped by the stands).

So we want to find the solution to this equation:

$$0 = -0.00132x^2 + 0.659x + 4.1$$

In Lesson 4-8, we learned how to solve a quadratic equation if the side with the variable is factored. But that's not the case here, and we don't know how to factor something with decimals. So we need a method to solve a quadratic equation that isn't in factored form. Fortunately, there's a formula that provides solutions to any quadratic equation, as long as it's written in the **standard form** $ax^2 + bx + c = 0$, where a is not zero.

Math Note

If you want to learn more about the math behind estimating home run distance, do a Google search for "how hit tracker works."

The Quadratic Formula

The solutions to a quadratic equation in the form $ax^2 + bx + c = 0$ are

$$x = \frac{-b + \sqrt{b^2 - 4ac}}{2a} \text{ and } x = \frac{-b - \sqrt{b^2 - 4ac}}{2a}$$

This is sometimes written more concisely as

$$x = \frac{-b \pm \sqrt{b^2 - 4ac}}{2a}$$

where the symbol "±" is read "plus or minus" and means one solution comes from adding and the other from subtracting.

Important Note

a is the coefficient of the term with the variable squared, *b* is the coefficient of the variable to the first, and *c* is the constant term.

1. For the equation describing Adam Dunn's home run on the previous page, what numbers correspond to a, b, and c in the quadratic formula?

2. Use the quadratic formula to find the two solutions to that equation. Round to one decimal place if necessary.

3. One of the solutions provides the length of the home run. What was it?

4. What is the significance of the other solution in the physical situation?

If we rewrite the two solutions provided by the quadratic formula just a bit, something interesting happens:

$$x = -\frac{b}{2a} + \frac{\sqrt{b^2 - 4ac}}{2a} \text{ and } x = -\frac{b}{2a} - \frac{\sqrt{b^2 - 4ac}}{2a}$$

Let's visualize that a bit differently:

$$x = -\frac{b}{2a} + \textit{some number} \text{ and } x = -\frac{b}{2a} - \textit{that same number}$$

This allows us to see that the x value $-\frac{b}{2a}$ is exactly halfway between the two solutions.

5. What's the connection between the solutions of the equation $ax^2 + bx + c = 0$ and the x intercepts of the graph of $f(x) = ax^2 + bx + c$?

6. The point where a parabola changes direction is called the **vertex** of the parabola. How do we know that the vertex has to be halfway between the x intercepts?

Combining Question 6 with the observation at the bottom of the previous page, we get a simple way to find the vertex of a parabola when we know the equation:

The Vertex Formula

For a quadratic function $f(x) = ax^2 + bx + c$, the x coordinate of the vertex can be found using the formula

$$x = -\frac{b}{2a}$$

Once you know the x coordinate, you can find the y coordinate by substituting the x coordinate in for x in the function.

The procedure we learned for finding a line of best fit using a graphing calculator back in Lesson 3-6 can also be used to find a quadratic function of best fit for a data set. Data is entered in the same way: the only difference is you choose **QuadReg** (choice 5) from the **STAT-CALC** menu rather than **LinReg** (choice 4).

Feet from home plate	Feet in the air
0	3
100	115
200	160
300	153

7. Write the quadratic function of best fit.

8. How high did the ball go?

9. How far did it go?

10. Look at the graph of your function compared to a scatter plot. How well do you think the equation fits?

4-9 Group

Experiment: The Time Needed To Drain a Bottle

Supplies needed:

- A clear 2-liter plastic bottle (or larger) with a hole drilled near the bottom of the straight-walled portion. The heights in the table below should be marked on the bottle either with a permanent marker, or by attaching a paper strip with height markings. Height zero should be marked at the height of the hole. Experiment with the hole size so that it takes roughly three minutes for the bottle to drain. About 1/4 inch is a good place to start.
- Water and a bucket to drain water into.
- A watch that displays seconds. The stop watch feature on a smartphone works well.

1. Cover the drain hole, then fill the bottle up to the 20 cm mark. Set the time to zero seconds. Then uncover the drain hole and take the cap off the bottle. As the water drains out, record the time for each height in the table.

x Time (sec)	y Height (cm)
0	20
	18
	16
	14

x Time (sec)	y Height (cm)
	12
	10
	8
	6

x Time (sec)	y Height (cm)
	4
	2
	0

2. Use your data to create a scatter plot on the grid provided.

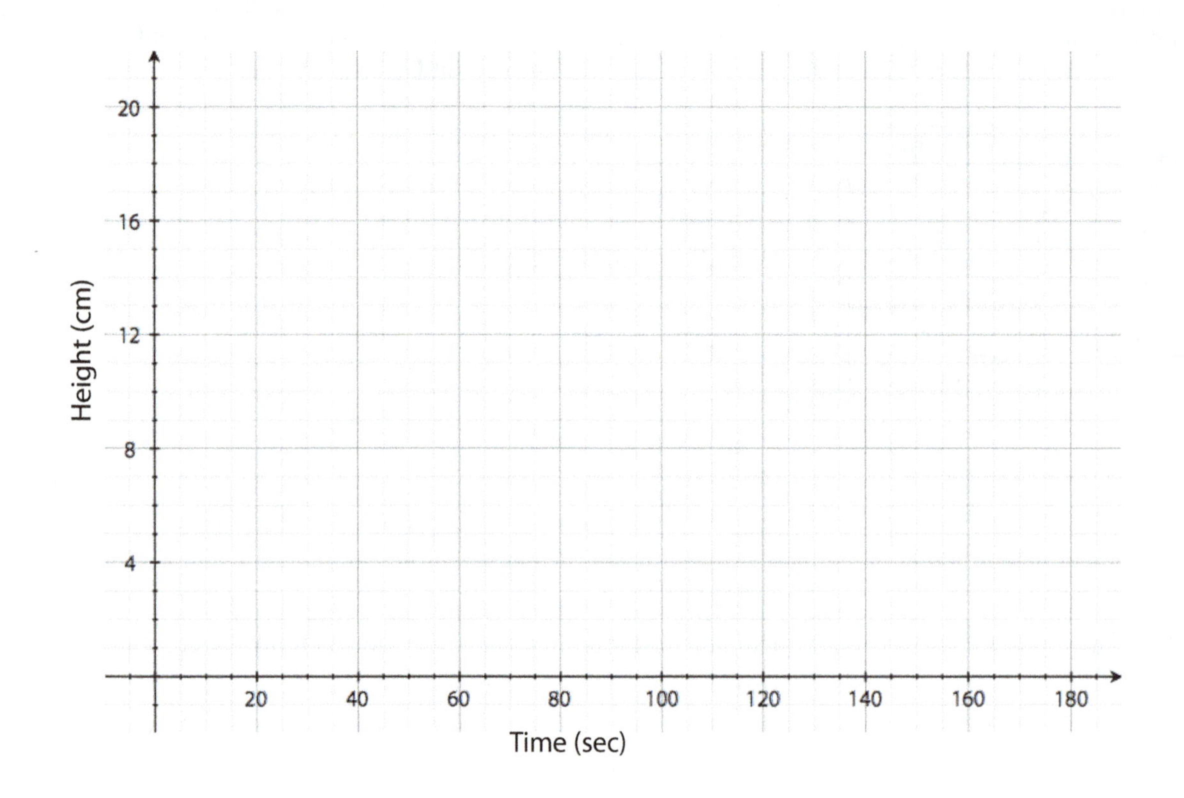

3. Do you think a linear or quadratic model is best suited to fit the data from this experiment?

4. What information in the table or graph supports your choice in Question 3? Explain why you think your choice makes sense.

5. Now try your best to ignore the scatter plot and data. Was there anything about what physically happened during the experiment that you think supports your choice in Question 3? Explain.

6. Find either the line or parabola of best fit, depending on which you chose in Question 3. You might encounter coefficients in scientific notation, so watch out for that. Round to 4 decimal places.

7. Use your model to predict the time when the water in the bottle reaches 2 cm above the drain hole. How does this compare to your experimental data?

8. What's the value of y when x is 50? Explain the meaning of these values.

9. Find the value of x when y is 11. Explain what these values mean, and compare to what you would predict based strictly on the table.

If you have time, repeat the experiment, this time recording the height every ten seconds. Make a table to record the results until the height reaches the drain hole. Then find a best-fit model again, and discuss how it compares to the model you found in Question 6.

4-9 Portfolio

Name ______________________________

Check each box when you've completed the task. Remember that your instructor will want you to turn in the portfolio pages you create.

Technology

1. ☐ Build a spreadsheet that finds the solutions of any quadratic equation of the form $ax^2 + bx + c = 0$ when values for a, b, and c are entered. Then use your spreadsheet to confirm the distances you found for the home run described on the first page of this lesson. Do the results agree with your answers to Question 2 in the class portion of this lesson? A template to help you get started can be found in the online resources for this lesson.

	A	B	C	D	E
1	a	b	c	Solution 1	Solution 2
2					

Skills

1. ☐ Include any written work from the online skills assignment along with any notes or questions about this lesson's content.

Applications

1. ☐ Complete the applications problems.

Reflections

Type a short answer to each question.

1. ☐ What is the quadratic formula used for? Why is that so useful? Discuss.
2. ☐ What types of applied problems can be solved using the vertex formula?
3. ☐ Take another look at your answer to Question 0 at the beginning of this lesson. Would you change your answer now that you've completed the lesson? How would you summarize the topic of this lesson now?
4. ☐ What questions do you have about this lesson?

Looking Ahead

1. ☐ Read the opening paragraph in Lesson 4-10 carefully, then answer Question 0 in preparation for that lesson.

4-9 Applications Name ______________________

The height of a golf ball in meters can be described by the equation $h = -4.9t^2 + 22.7t$, where t is the number of seconds after it was hit. (Note: This is different than our home run equations, where the input x represented the number of horizontal feet traveled.)

1. Use a calculator or spreadsheet to make a table of inputs and outputs for this function. The variable t should start out at zero and increment by 0.2. Then use the result to draw a detailed graph of the function.

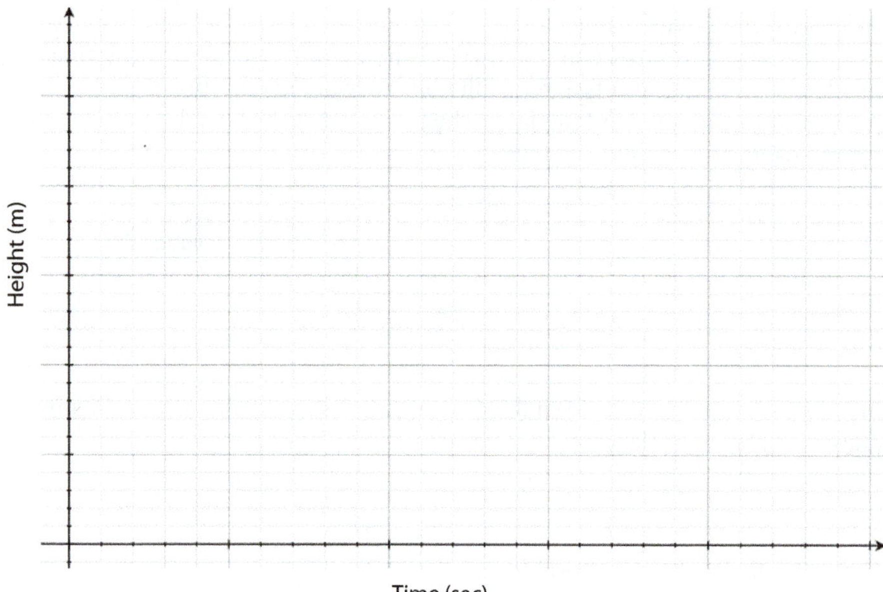

2. Find the vertex of the parabola, rounding each coordinate to the nearest tenth.

3. Explain what each coordinate of the vertex means.

4. Find the intercepts for the function.

5. Explain what each intercept means.

4-9 Applications Name ____________________

6. Describe when the ball is headed upward, and when it's coming back down.

7. The function $d(t) = 48.6t$ describes the horizontal distance (in meters) traveled by the ball after t seconds. How far did it go?

8. The distance for golf shots is traditionally measured in yards. Given that there are 3 feet in a yard and that 1 meter is 3.28 feet, how far in yards did the shot go?

9. The table shows data describing yet another home run. Find a quadratic function of best fit for the data. See Question 7 in the Class portion of this lesson for a refresher.

Feet from home plate	Feet in the air
0	4
100	75
200	95
300	46

10. How far away from home plate did the ball land?

11. How high did it go?

12. If the ball went over the fence 340 feet from home plate, and the fence is ten feet high, by how much did the ball clear the fence?

Lesson 4-10 Follow the Bouncing Golf Ball

Learning Objectives

- ☐ 1. Find an exponential curve of best fit for data.
- ☐ 2. Study the decay rate for exponential decay.

It's just a glorious day. The only way to ruin a day like this would be to play golf on it.
– David Feherty

Avid golfers have a classic love-hate relationship with the game. They enjoy it enough to spend a lot of time and money pursuing it, but they also know how downright maddening it can be. Golf is a whole lot harder than it looks, and sometimes it seems like the ball just moves and bounces with a mind of its own–an evil mind intent on ruining the golfer's day.

Even so, like everything else in our world, golf balls are bound by the laws of physics and their motion actually is predictable–at least in a lab setting, if not on the golf course. In this lesson, we'll follow a bouncing golf ball to help us complete our study of exponential growth and decay, relying on technology to find exponential equations that best fit a data set.

0. After reading the opening paragraph, what do you think the main topic of this section will be?

4-10 Group

Here's the list of supplies needed for our golf ball experiment:

1. A golf ball; without this, it's not a golf ball experiment, is it?
2. A meter stick or tape measure with metric units
3. A chair or short stepladder
4. Masking tape and a marking pen
5. A hard, smooth floor area next to a wall
6. A graphing calculator with a regression feature

Part 1: The Bounce Lab

Location is everything here: you'll need a relatively high ceiling, and a smooth concrete or other hard floor. Tile with grout lines won't work: the grout lines will affect the bounces of the ball. The key is to get consistent bounces straight up with no deflection. Have the tallest member of your group stand on a chair or short stepladder, and run a strip of masking tape vertically along the wall from the floor to the highest point they can reach. (Ideally, the starting point will be at least 250 cm above the floor.) Using your meter stick/tape measure and the marking pen, label heights on the masking tape every 10 cm starting with 0 cm at floor level.

Now you're ready to bounce. Hold the golf ball a few inches from the wall with the bottom of the ball at the highest height marked on the tape. Drop the ball and have someone in the group note the approximate height the ball reaches at the top of its first bounce. This will give you a rough idea of where to look. Now repeat the drop, and have the person that watched the height of the first bounce put their finger on the tape at the height the ball reaches on the first bounce (the bottom of the ball, that is). Then mark that height with the pen. Repeat twice more, then calculate and record the average of the three bounce heights. (Use the table on the next page to record your results.)

Next, we'd like to calculate the height of the second bounce. But rather than having the ball bounce twice, use the height of the first bounce that you calculated as the new release point. Clever, right? Repeat three times, recording the bounce heights in the table and calculating the average. Then use that height as release point to calculate the height of the third bounce, and so on. Continue until you have heights for the first 8-10 bounces.

1. Record the data from your experiment in the table.

x Bounce #	Trial 1 bounce height (cm)	Trial 2 bounce height (cm)	Trial 3 bounce height (cm)	Total bounce height (cm)	y Average bounce height (cm)
0					
1					
2					
3					
4					
5					
6					
7					
8					
9					
10					

2. Use this data to create a scatter plot. Use the average bounce height as your y coordinates.

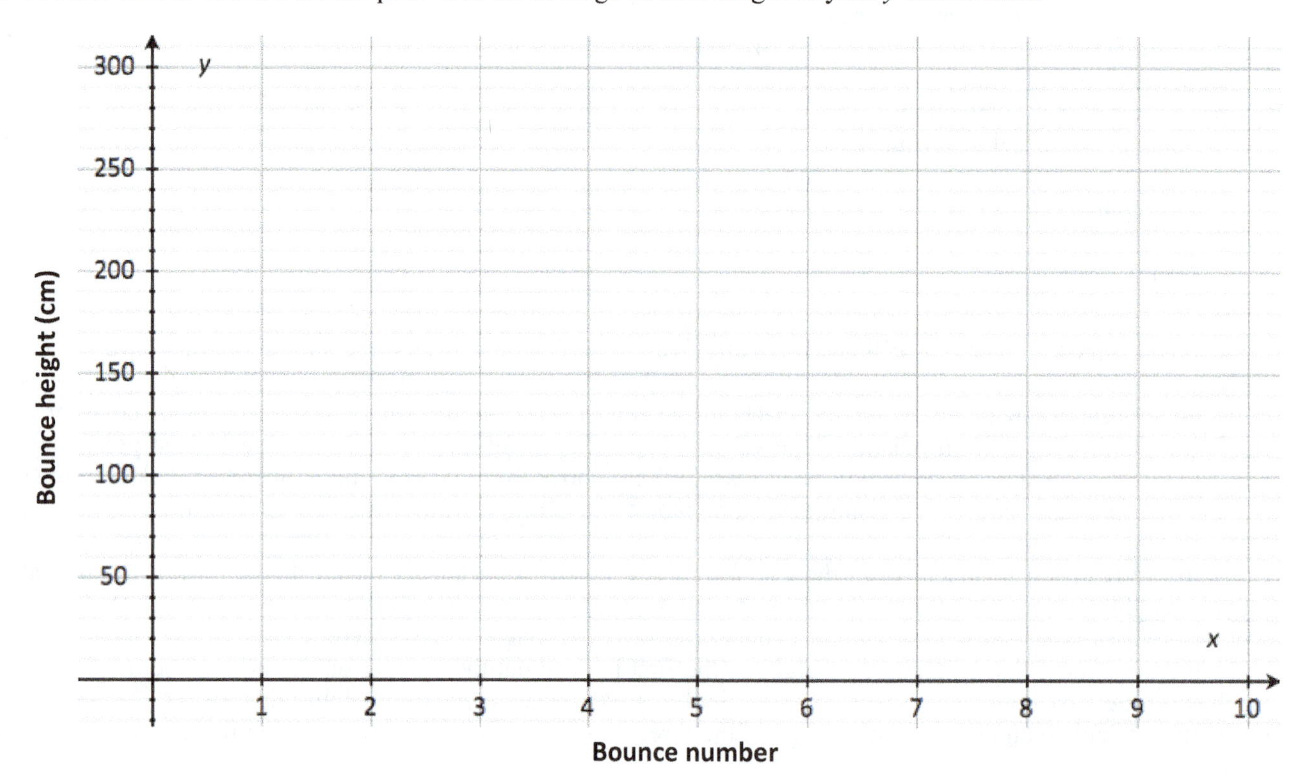

Part 2: Modeling Your Data

3. Using your graphing calculator, find an exponential equation of best fit for your data. The command for this is usually **ExpReg** on the **STAT-CALC** menu; your equation should be of the form $y = ab^x$. Round to two decimal places.

$y =$ ______________________

4. Use your equation to approximate the height of the 7th bounce. How does the result compare to the value in your table? Does this surprise you?

5. Evaluate your equation for $x = 0$. What does your answer mean?

Now let's analyze the significance of every symbol in our equation.

6. What does the variable x represent?

7. In the equation $y = ab^x$ that models your data, what is the value of a? What does it tell us about the experiment? Your answers to 5 and 6 should help.

8. In the equation $y = ab^x$ that models your data, what is the value of b? What does it tell us about the experiment?

Part 3: The Connection Between Exponential Decay and the Rate of Decay

In at least one lesson in each unit of this book, we've studied exponential growth or decay. The independent variables we've worked with have typically been whole numbers: years after an account was opened, hours since caffeine was introduced to the bloodstream, number of bounces for a golf ball, etc. In each case, the distinguishing feature is that each new output comes from multiplying the previous by the same fixed number. In the bounce lab, we see that same pattern: each bounce is a certain percentage of the height the ball was dropped from. The result is an equation where the input is an exponent, and that's why we call this type of function an **exponential function.**

Now let's study the rate of decay.

9. Using data from your original table, fill in this new table. The drop height corresponds to the height of the previous bounce; start with the original height the ball was dropped from. Use the average bounce heights from the original table

x Drop height (cm)	y Bounce height (cm)

x Drop height (cm)	y Bounce height (cm)

10. Use this data to create a second scatter plot.

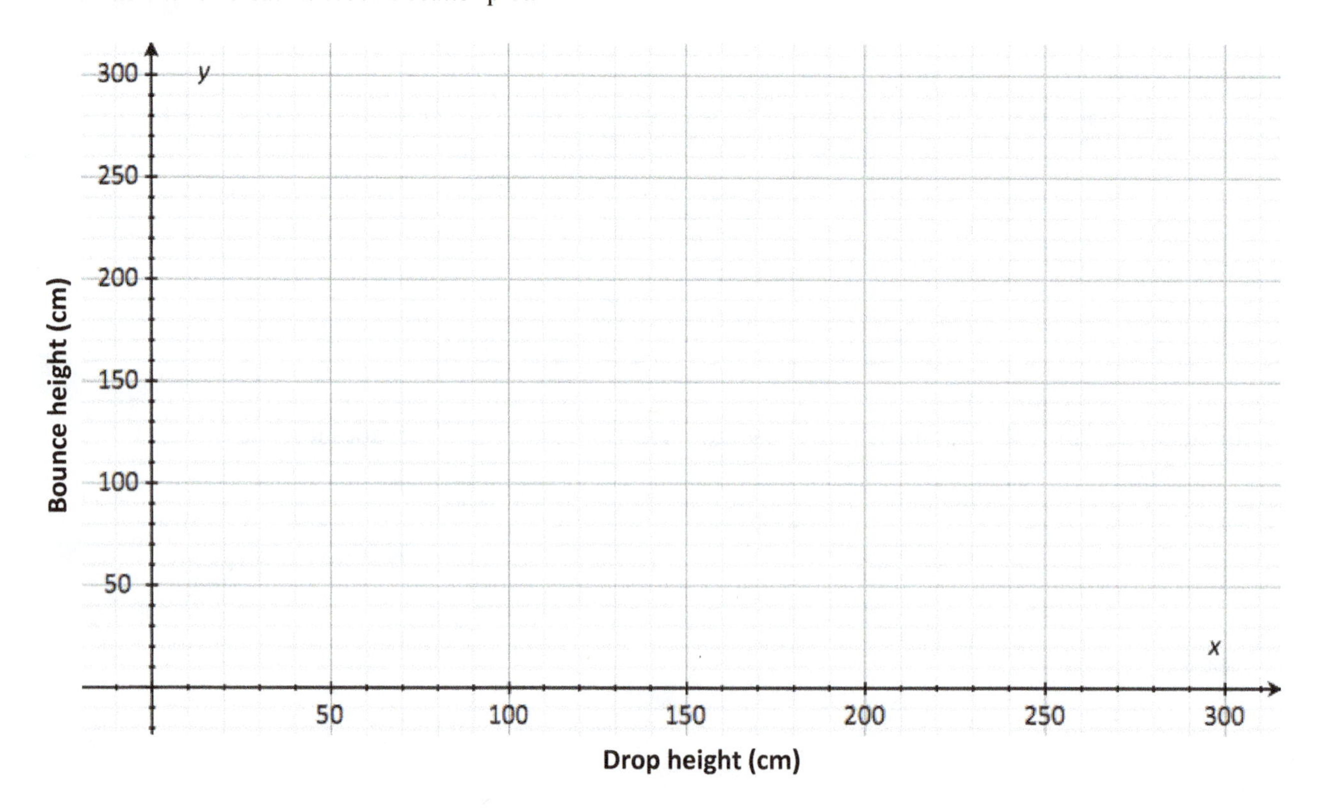

11. Based on the scatter plot, what kind of equation seems to be a good choice for modeling this data?

12. Use your graphing calculator to find the equation of best fit. Round to two decimal places.

$y =$ ___________________________

13. Your equation has a variable in it, but also has **parameters:** these are constants in an equation of a certain form that are determined by the data for a particular model. For example, when you found an equation of the form $y = ab^x$ earlier, x is the variable, while the values you found for a and b are parameters. Identify the variables and parameters in your answer to Question 12, the describe what each means in this setting.

14. If you were to drop the golf ball from your experiment out of a second story window 20 feet above a concrete patio, how high would you expect it to bounce?

15. Find the relative change in height from one bounce to the next. If you need a reminder, we studied relative change in Lesson 2-4.

16. (This one requires some thought, but is the key question.) How does your answer to Question 15 relate to EACH of the best-fit equations found in this lesson?

According to the International Federation of Tennis, a regulation tennis ball should bounce about 55 inches when dropped from a height of 100 inches.

17. What's the bounce height as a percentage of the drop height for a regulation tennis ball?

18. Write an expression that calculates the bounce height for a regulation tennis ball dropped from 200 inches.

19. Write an equation that describes the bounce height (y) of a regulation tennis ball when it's dropped from x inches.

20. Using your answer to Question 17, it's possible to write an equation that describes the height of the xth bounce when a regulation tennis ball is dropped from a height of 100 inches. Do it! The things we learned in the golf ball experiment will definitely come in handy.

21. What do you think would be some sources of error in the golf ball experiment? That is, what factors might cause inaccuracy?

4-10 Portfolio

Name ____________________

Check each box when you've completed the task. Remember that your instructor will want you to turn in the portfolio pages you create.

Technology

1. ☐ Questions 1, 2, and 3 in the Group section are great examples of the three ways we've modeled data: numerically, graphically, and algebraically. Build an Excel spreadsheet that illustrates each of those models, using a table, connected scatter plot, and a formula. A template to help you get started is available in the online resources for this lesson. (Note: The exponential regression equation supplied by Excel will be in a different form than the one you got from your graphing calculator. The Excel version uses the number e as its base: this number is about 2.7. Make a table of values and a graph for each of the two equations to verify that they're equivalent.)

Skills

1. ☐ Include any written work from the online skills assignment along with any notes or questions about this lesson's content.

Applications

1. ☐ Complete the applications problems.

Reflections

Type a short answer to each question.

1. ☐ We've studied linear, quadratic, and exponential lines of best fit in this course. If you're looking at a set of data, how can you decide which of those three types of model would be most appropriate for the data?
2. ☐ How could you tell if none of those three types of equations is a good fit for a data set?
3. ☐ Take another look at your answer to Question 0 at the beginning of this lesson. Would you change your answer now that you've completed the lesson? How would you summarize the topic of this lesson now?
4. ☐ What questions do you have about this lesson?

Looking Ahead

1. ☐ Start preparing for your final exam–good luck!

4-10 Applications Name ____________________

1. The equation $y = 10{,}000(1.06)^x$ describes the growth of an investment based on the number of years it's been allowed to grow. Identify the variables and parameters in this equation, and describe what the significance of each is in investment terms.

2. Use the equation in Question 1 to complete the table, then use the table to draw a graph of the equation. Don't forget to label an appropriate scale on each axis.

Years	Value
0	
5	
10	
15	
20	

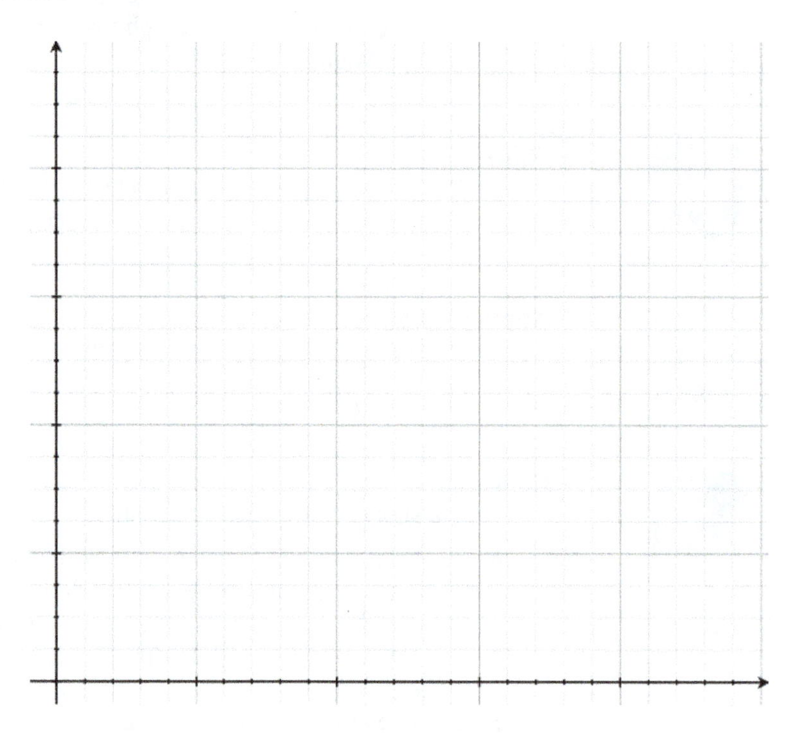

3. The equation $y = 10{,}000(1.06)^x$ illustrates ____________________ growth. The value of the investment was growing by _______% each year.

4-10 Applications Name ____________________

The model in Questions 1-3 is an example of interest **compounded annually.** This means that the full 6% of interest is added to the account at the end of one year. This doesn't sound very fair to someone that invests their money for 11 months–they get no interest at all. This became a competitive disadvantage for financial institutions, and some began to divide the annual interest into periodic shares, so that (for example) you could get 1/12th of that 6% each month. When this happens, we say that interest is **compounded monthly.** Interest can also be compounded weekly (52 times per year), quarterly (4 times per year), daily (365 times per year), or really any other period you could think of.

4. If interest is compounded monthly, what growth factor would be needed to provide 1/12th of 6% interest each month?

5. How many times would this growth factor be applied in 2 years? In 5 years? In t years?

When an initial amount of P dollars is invested at r% annual interest compounded m times per year, the value of the account (A) after t years is given by the equation

$$A = P\left(1+\frac{r}{m}\right)^{mt}$$

6. Describe the significance of $\left(1+\frac{r}{m}\right)$ in the formula. (Think about Question 4.)

7. Describe the significance of mt in the formula. (Think about Question 5.)

8. Write an equation that represents the value in an account that starts out with an initial investment of $10,000 and pays 6% interest compounded monthly.

9. Use the equation in Question 8 to complete the table, then use the table to draw a graph of the equation. Don't forget to label an appropriate scale on each axis.

Years (t)	Value (A)
0	
5	
10	
15	
20	

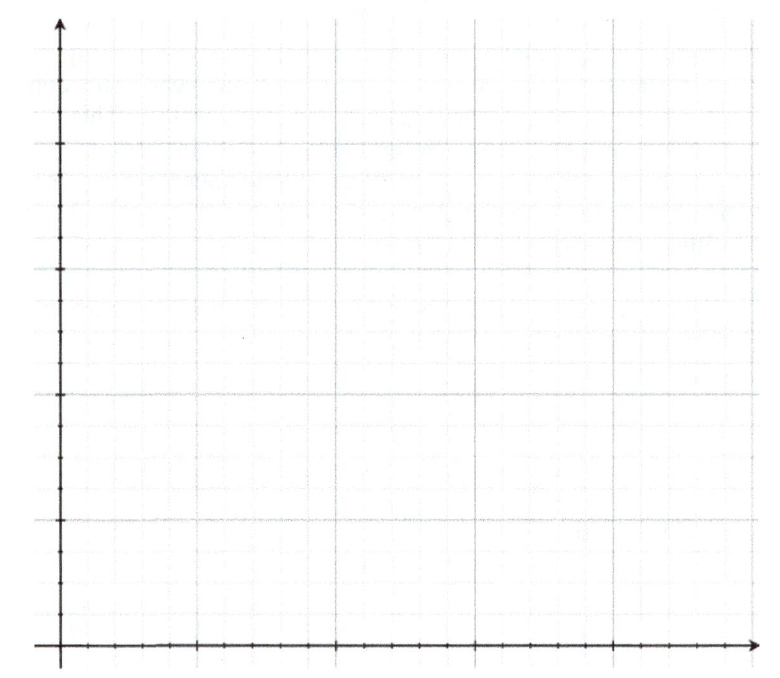

One of the best reasons for understanding exponential growth is that it will help you plan for your financial future. Let's dream a little: suppose that you win $100,000 in a lottery. Sure, you could blow it on silly stuff like food and shelter, but instead you choose to invest it and let it grow for 20 years. Smart! In 20 years, you look back to analyze the growth of your investment, and you find the values in the next table.

Years from now (x)	Value (y)
0	$100,000
5	$133,775
7	$150,440
13	$212,055
20	$324,410

10. Use your graphing calculator to find the exponential equation of best fit for the data. Your equation should be in the form $y = ab^x$.

11. Describe the meaning of the parameters a and b in your equation.

12. Use your equation to estimate the value of the investment in 30 years.

13. Graph the equation with a graphing calculator and use the graph to estimate how long it would take for the investment to grow to $500,000.

Trig Supplement The Importance of Accessibility

Learning Objectives

- ☐ 1. Study similar triangles.
- ☐ 2. Define the trigonometric ratios.
- ☐ 3. Use inverse trigonometric commands.
- ☐ 4. Use trigonometric ratios to solve problems.

The role of genius is not to complicate the simple, but to simplify the complicated
– Criss Jami

Most of us have gotten pretty accustomed to seeing a variety of accessibility options for people with disabilities. So it seems hard to believe that the Americans with Disabilities Act, which was instrumental in improving access for folks with a wide range of disabilities wasn't enacted until 1990. Consider entrance ramps, which are now required by law for all buildings used for public or business purposes. Not surprisingly, there's more to access ramps than just throwing a piece of plywood over stairs: a ramp that's too steep not only prevents access, it presents a major safety hazard. In fact, the ADA requires an angle between an access ramp and the ground to be at most 4.76°. Where in the world does such an odd number come from? In this lesson, we'll study a method of working with triangles that has applications in engineering, navigation, building, home improvement, and many other areas, including accessibility.

0. After reading the opening paragraph, what do you think the main topic of this section will be?

Trig Class

Trigonometry is a subject in which we study the relationships between the measures of sides and angles in a triangle. The goal is to be able to solve geometric problems that involve triangle diagrams, like ones involving access ramps. To begin, we'll study a related idea, **similar triangles**. Triangles are similar when they have the exact same shape, but different sizes.

1. An access ramp that covers a vertical incline of 20 inches has to cover at least 240 inches horizontally. If the vertical incline is 10 inches, how much horizontal distance is needed?

How much horizontal distance would be needed for a vertical incline of 40 inches?

Notice that in your answers to Questions 1 and 2, the ratio of the horizontal distance to the vertical incline is always the same:

$$\frac{240\text{ inches}}{20\text{ inches}}=\frac{12}{1} \qquad \frac{120\text{ inches}}{10\text{ inches}}=\frac{12}{1} \qquad \frac{480\text{ inches}}{40\text{ inches}}=\frac{12}{1}$$

When the ratios of corresponding sides of two triangles are identical, then the two triangles have the same shape, and we call them **similar triangles**. If we look at a diagram of the three ramps in Questions 1 and 2, we can see that they all have the same shape:

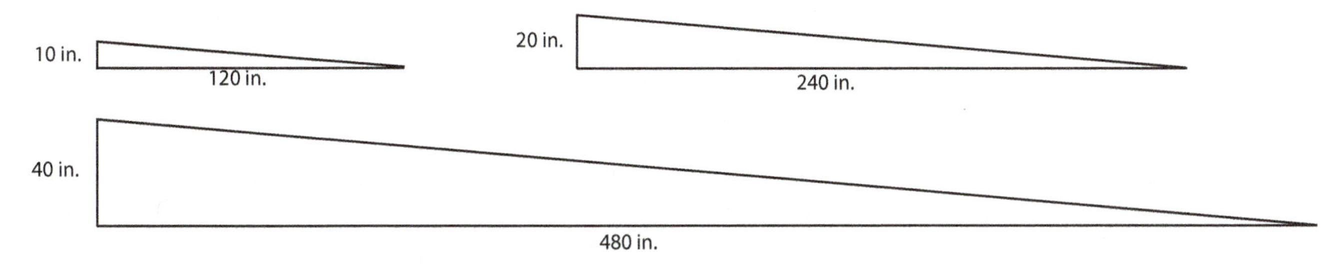

is gives them all the same angle of incline, which is good since that was pretty much the point. Because the ios of side lengths is a perfectly good way to describe the shape of a triangle, those ratios are given special names for right triangles (which we know are triangles that have a 90° angle). The names given to the ratios of the sides in a right triangle are called **trigonometric ratios**, and there are three of them: **sine** (abbreviated sin), **cosine** (abbreviated cos), and **tangent** (abbreviated tan). These ratios are defined as follows. (Note how the angles and side lengths of the triangle are described in the accompanying diagram.)

The Trigonometric Ratios

$$\sin(A)=\frac{\text{Length of side opposite angle }A}{\text{Length of hypotenuse}}=\frac{a}{c}$$

$$\cos(A)=\frac{\text{Length of side adjacent to angle }A}{\text{Length of hypotenuse}}=\frac{b}{c}$$

$$\tan(A)=\frac{\text{Length of side opposite angle }A}{\text{Length of side adjacent to angle }A}=\frac{a}{b}$$

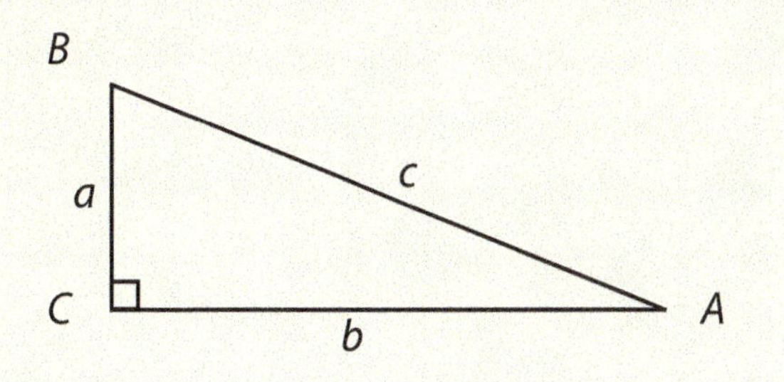

3. What is the tangent of the angle of incline for the access ramp triangles above?

Math Note

Recall that the side across from the right angle in a triangle is called the **hypotenuse**.

4. What are the sine and the cosine of the angle of incline for the access ramp triangles?

Values for the trigonometric functions can be computed using your calculator: most graphing calculators have buttons marked SIN, COS, and TAN for this purpose. This allows us to solve problems involving right triangles when we know the measure of an angle.

Notice that we used the term "trigonometric functions" in the above paragraph. We learned about the general function symbol $f(x)$ in Lesson 4-8. With trig ratios, we're dealing with very specific functions that are designed to perform a specific task, so rather than use the generic $f(x)$, we use specific symbols to indicate what each function does. In the case of sine, for example, we use the symbol $\sin(A)$ to represent the sine of angle A. The letters "sin" are taking the place of f in standard function notation, and angle A is taking the place of the standard input x. You can think of all of "sin", "cos", and "tan" as in *instruction* to do something to the angle that follows: $\tan(A)$ tells us to take input angle A in a right triangle and output the ratio of the length of the side opposite to the length of the side adjacent. (By the way, this means that you should NEVER write of the trig function symbols with nothing after it. The symbol sin doesn't mean anything unless it has an angle after it!)

Now back to using calculators to find outputs for the trig functions. For example, the safety recommendation for one 9 foot ladder is that it should be placed at no steeper than a 65° angle with the ground. It would be perfectly reasonable to want to know how far away from the base of a wall you can safely put the ladder, and trigonometry can help us do that.

Math Note

Remember, the function symbol $f(x)$ is read "f of x", not "f times x". So we read $\sin(A)$, $\cos(A)$ and $\tan(A)$ as "sine of A", "cosine of A" and "tangent of A" respectively.

5. Draw a diagram that shows the ladder placed at a 65° angle with the ground, leaning against a vertical wall. Label the angle you know and the length of the ladder, and use the variable x to represent the distance between the bottom of the ladder and the wall.

6. Which of the trig ratios involves the length of the ladder and the side labeled with x in your diagram?

7. Use the trig ratio from Question 6 to set up an equation involving variable x.

8. Using a calculator to get a value for the trig ratio involved in your equation (see Using Technology below), solve your equation to find how far away from the wall the bottom of the ladder has to be placed.

Using Technology: Computing Values for Trig Ratios

TI-84 Plus Calculator

1. Make sure that your calculator is in degree mode: press MODE, then use the arrow and enter keys to select DEGREE.
2. Press the key corresponding to the trig ratio you need a value for, either SIN, COS, or TAN.
3. Enter the degree measure of the angle, then press) and ENTER.

Excel Spreadsheet

1. To compute sine of an angle measured in degrees, choose a blank cell and enter "=SIN(RADIANS(angle))", where angle is the degree measure of the angle.
2. To compute cosine of an angle measured in degrees, choose a blank cell and enter "=COS(RADIANS(angle))", where angle is the degree measure of the angle.
3. To compute tangent of an angle measured in degrees, choose a blank cell and enter "=TAN(RADIANS(angle))", where angle is the degree measure of the angle.

Note: The "RADIANS" command converts the angle in degree measure to radians, which is a different unit for measuring angles.

If you look closely at the SIN key on a graphing calculator, you'll see the symbol SIN^{-1} above it. This is a built-in feature known as **inverse sine** that allows us to find the measure of an angle when we know the sine ratio that it corresponds to. Similarly, there are **inverse cosine** (COS^{-1}) and **inverse tangent** (TAN^{-1}) commands that do the same for the other trig ratios.

For example, if we know that $\sin(A) = 0.3$, we can find the measure of angle A by using a calculator to find $\sin^{-1}(0.3)$. The string below shows how to compute inverse sine of 0.3 on a TI-84. (Make sure the calculator is in degree mode, as describe in the above tech box.)

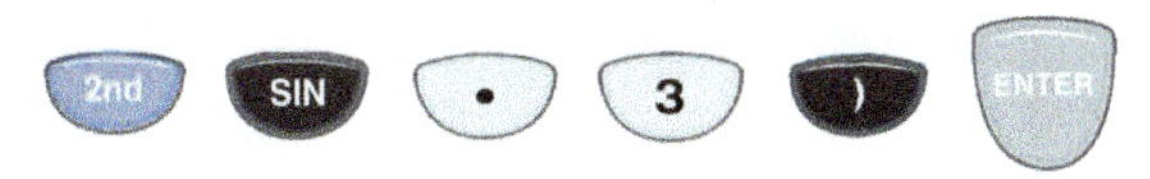

9. If $\cos(B) = 0.74$, what is the degree measure of angle B?

10. Use the access ramp triangles following Question 2 to find the incline angle for the ramp illustrated. Where have we seen the answer before?

11. Use trig ratios to find the angle at the top of the ramp in the triangles following Question 2.

12. How could you have found the measure of that angle without trig ratios? (Hint: Your answer to Question 10 would surely help.)

Trig Group

One of the most important aspects of science is experimenting to verify theoretical results. So let's see how well our trig ratios predict the various measurements in triangles. For this part of the activity, you'll need a ruler and a protractor (and yes, there are apps for that!)

1. Use the ruler to measure the legs of each right triangle below in millimeters. (Recall that the legs are the sides that come together to form the right angle.) Label these measurements on each triangle.

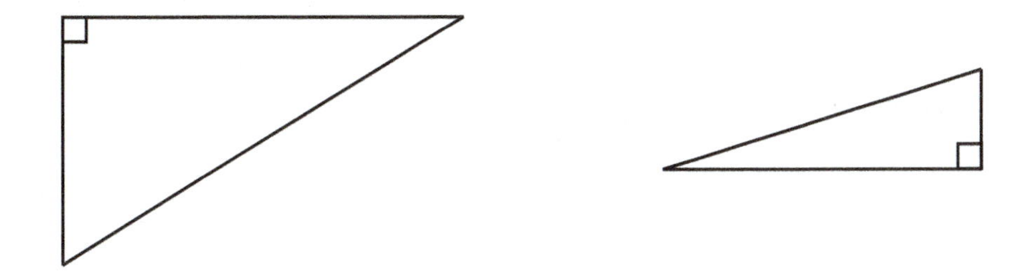

2. Use the Pythagorean theorem to calculate the length of each hypotenuse, then label those lengths on each triangle.

3. Measure each hypotenuse using the ruler. How accurate was your calculation in Question 2? If it's not exact, what do you think is most responsible for the error?

4. Using the tangent ratio, write an equation that will calculate the measure of the bottom angle in the first triangle, then use the inverse tangent command on a calculator to calculate the angle.

5. Measure that angle with your protractor. How close are the results to what you calculated in Question 4?

6. Measure the angle at the bottom left side of the second triangle with your protractor.

7. Repeat the procedure in Question 4 to calculate the measure of the angle you just measured, but use either the sine or cosine ratio. (You can decide which.) Describe how accurate you think your result is, and what might account for any error.

This part of the activity will require a tape measure. Find an access ramp for entering the building you're in. If there isn't one and you're on campus, report your school to the government for violating the Americans with Disabilities Act, and find a nearby building that does have an access ramp.

8. Measure the length of the ramp, placing the tape measure along the ramp. Then measure either the horizontal run or the height, or both if you like. (You may need a little ingenuity to get an accurate measurement here.) Next, draw a triangle diagram with your measurements below.

Use a trig ratio to calculate the angle of incline. Is it within the guidelines specified by the ADA?

10. How much would the length of the ramp change if the angle exactly matched the maximum incline specified by the ADA?

Trig Portfolio

Name ______________________________

Check each box when you've completed the task. Remember that your instructor will want you to turn in the portfolio pages you create.

Technology

1. ☐ We measured angles in degrees in this lesson, but mentioned another unit for measuring angles, the radian. Using the Internet as a resource, look up radian measure. Write a description of what radian measure is based on, then write a unit fraction that can be used to convert degrees to radians, and one that can be used to convert radians to degrees.
2. ☐ Use your conversion factors to build a spreadsheet that converts angles in degree measure to radian measure, and angles in radian measure to degree measure. Use it to convert 10, 20, 30, 40, 50, 60, 70, 80, and 90 degrees to radian measure, and to convert 1, 2, 3, 4, 5 and 6 radians to degree measure.

Skills

1. ☐ Include any written work from the online skills assignment along with any notes or questions about this lesson's content.

Applications

1. ☐ Complete the applications problems.

Reflections

Type a short answer to each question.

1. ☐ What is a trigonometric ratio?
2. ☐ List some of the things that trigonometric ratios can be used for.
3. ☐ Take another look at your answer to Question 0 at the beginning of this lesson. Would you change your answer now that you've completed the lesson? How would you summarize the topic of this lesson now?
4. ☐ What questions do you have about this lesson?

Looking Ahead

1. ☐ Read the opening paragraph in the next lesson assigned by your instructor carefully, then answer Question 0 in preparation for that lesson.

Trig Applications Name ______________________________

1. One common use of trig ratios is measuring inaccessible objects. Unless you can walk on water, it's pretty hard to measure the distance across a lake, but with some basic measurements and a knowledge of trig, it can be done. How wide is the lake in the diagram?

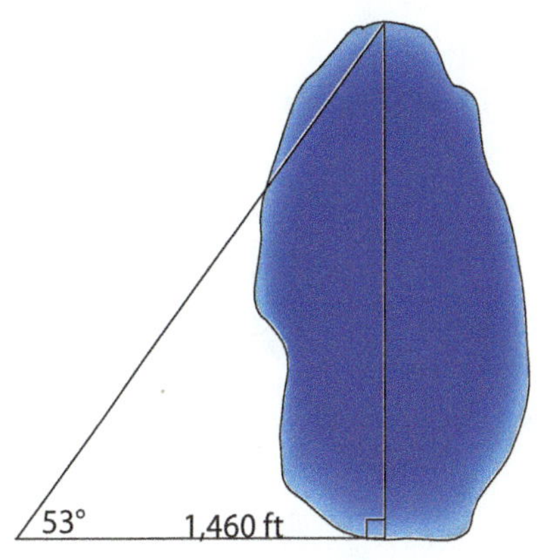

2. When I was finishing my basement, I needed to cut a length of baseboard to run along the stairs. The hard part was guessing what angle to cut on the bottom so the cut was parallel to the floor. So I didn't guess: I used trigonometry! (It turns out that I'm a math professor.) Taking some simple measurements, I found that the stairs covered a horizontal span of 12 feet 3 inches, and went down by 8 feet 6 inches. What angle did I need to cut at the bottom? Include a diagram and a description of what exact angle you found.

3. A hotel security camera is mounted 9 feet 6 inches above floor level at the end of a hallway. The installation guide says that in order to get the widest field of vision, the camera should be pointed at a spot 24 feet down the hall. At what angle should the camera be aimed downward? Again, a diagram and an explanation of what angle you found seems like a splendid idea.

Trig Applications

Name ______________________________

The Great Pyramid at Giza is one of the oldest and largest of the great pyramids of Egypt. It was constructed over 4,500 years ago, making it one of the most remarkable feats of engineering in human history. Needless to say, it would be pretty much impossible to measure the height of the pyramid directly. That's why trigonometry was originally used to calculate the height. The diagram below is a scale model of the faces of the pyramid: the scale is 1: 2,000. (Recall that this means that 1 in. on the diagram corresponds to 2,000 in. on the actual pyramid.) Let's simulate the original height measurement using our diagram.

4. The easiest measurement to take on the actual pyramid is the length of the base. Use a ruler to measure the base on our scale model, then use the scale and a conversion factor to find the length of the actual base in feet, and mark that on the diagram.

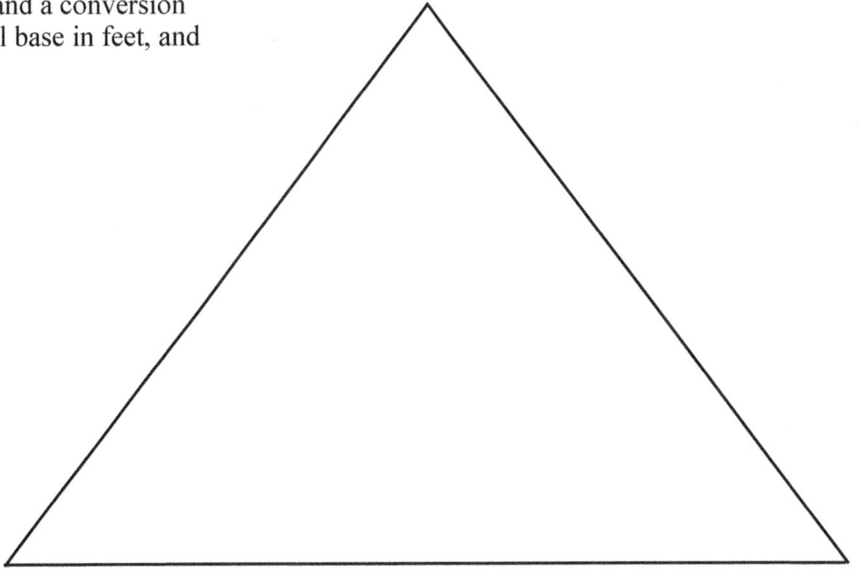

5. It also wouldn't be hard to measure the angles at ground level. Measure those angles with a protractor (they're both the same) and mark the angle on the diagram. (If you don't have a protractor, you can get a protractor app on a smartphone, or print one from the Internet. That's pretty awesome.)

6. Use your measurements to calculate the height of the triangle in the diagram.

e height you found in Question 6 is not the actual height of the pyramid: it's the distance up one of the faces, rked F in the next diagram.

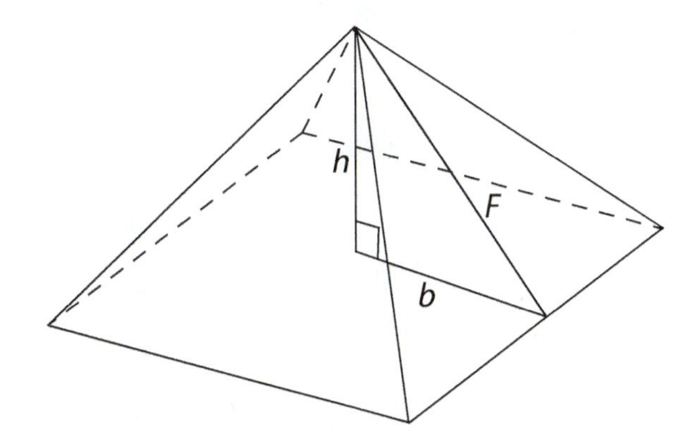

7. Redraw the right triangle whose sides are labeled in the diagram above, and label the hypotenuse with your answer from Question 6.

8. Label the side marked b on the above diagram on your triangle. (Hint: that side goes halfway across the base of the pyramid.)

9. Use the Pythagorean theorem to find the height of the Great Pyramid, then use the Internet as a resource to see how accurate your calculation is.

Stats Supplement The Error of Your Ways

Learning Objectives

- ☐ 1. Identify margin of error in a poll.
- ☐ 2. Calculate and interpret the meaning of margin of error.
- ☐ 3. Calculate the number of respondents needed for a given margin of error.

Better to trust the man who is frequently in error than the one who is never in doubt.
–Eric Sevareid

Do you know what percentage of Americans think that Bigfoot exists? Me neither. But I bet you can find that information online somewhere. In our information society, it seems like there's a poll for everything, from approval of the President to favorite type of underwear. And since so much information is provided to citizens and consumers in the form of polls, an understanding of how polls operate has become a pretty important survival skill. In this lesson, we'll study an important aspect of interpreting the results of a poll, the margin of error. And we'll find out that understanding error in polls can prevent errors in judgement.

0. After reading the opening paragraph, what do you think the main topic of this section will be?

Stats Group

Polling has become big business, and two of the biggest names in that business are Gallup and Pollster. One of the more widely-quoted polls they run is used to determine the President's current approval rating among citizens. But of course, they can't ask everyone in the country every week whether they approve or disapprove of the President's performance, so instead they ask a sample of citizens and use sampling and statistical techniques to try to decide if the results are truly indicative of the general population.

1. Go to the website www.pollster.com, then find the link for the presidential approval survey and click on it. What is the President's most current approval rating? Write a sentence explaining what that means.

- How would you describe the trend in approval rating over the last six months?

3. Scroll down a bit and find the first link to the Gallup survey and click on it. You'll see a brief description of how the poll was conducted, including a line that says "Margin of error", followed by something like ±3 percentage points." What do you think the term "margin of error" means in this context?

4. Can you think of another situation in life, school, or work where you've heard either "margin of error", or "margin for error"? Describe.

5. What do you think is better to have, a small margin of error or a big one? Does it depend on the situation?

Stats Class

Because it's not possible to poll every individual all the time to ask their opinions, sampling plays a really important role in polling. **Sampling** is the process of choosing a portion of a population in such a way that the individuals chosen are in some way representative of an entire population. The downside of sampling is that it introduces uncertainty in the results of a poll, since to some extent the pollster is making an educated guess.

The margin of error for a poll is a statistical measure that provides a range of values that the true outcome of the poll is likely to be inside. For example, if someone reports that 84% of Americans like pizza, and the margin of error is $\pm 3\%$, this means they're confident that the actual percentage of Americans that like pizza is somewhere between 81 and 87% (since those are the percentages that are within 3% of 84.)

Here's the catch: how confident is confident? For margins of error, confidence is measured as a percentage. One way of computing margin of error is with a 95% confidence level. In our example above, a 95% confidence level would mean that the polling institution is 95% sure that the accurate percentage of Americans that like pizza is between 81% and 87%.

1. An Associated Press poll indicated that 63% of teachers below the level of college feel like the amount of homework students get is not excessive. The poll had a margin of error of ±3.5% with a 95% confidence level. Write a sentence or two explaining what that result means.

A formula for computing margin of error is provided in the colored box below. Sadly, it means absolutely nothing until you know what all of the symbols stand for, which is why you should read everything in the colored box.

Margin of Error

When n people are polled, and $\hat{p}$ give a particular response, the margin of error with a 95% confidence interval is given by the formula

$$\text{Margin of Error} = 2\sqrt{\frac{\hat{p}(1-\hat{p})}{n}}$$

Note that both the margin of error and $\hat{p}$ (which is often called the **sample** proportion) are percentages written in decimal form.

The margin of error means we that if this poll were conducted many times, we would expect that at least 95% of the time the actual number of people in the population that would give a particular response is somewhere between $\hat{p}$ − the margin of error and $\hat{p}$ + the margin of error.

A Wall Street Journal poll conducted in December of 2014 asked 1,000 adult Americans if they or a family member had ever been notified by a credit card company of a possible data breach involving their personal information. Forty-five percent reported that they had been notified. Use the formula in the colored box to find the margin of error for this result.

3. Write a sentence describing exactly what your answer to Question 2 means about the overall percentage of American adults that have been notified of a data breach. More detail is always better.

4. For this question, don't look at the margin of error formula. Seriously, no peeking or I'll know. Based on what margin of error measures, do you think the margin of error for a poll should go up or down as the number of people surveyed gets larger? Justify your answer.

5. Now look at the margin of error formula. Based strictly on that formula, what should happen to the size of the margin of error if the number of people surveyed gets larger? Does this match your answer to Question 4? Explain.

6. What if the folks from the Wall Street Journal who conducted the data breach poll got a little bit lazy, and decided to only survey 200 people, then go bowling. Recalculate the margin of error. Did it get bigger or smaller?

While it's absolutely fine and dandy to calculate the margin of error after conducting a poll, in many cases it's more useful to think about what the margin of error might be like BEFORE you're done sampling. This would allow you to decide how many people need to be surveyed in order to make your result reliable.

Suppose that you begin surveying students on your campus about their attitudes on race relations in the U.S. After 70 surveys are returned, you find that 32% of respondents think that race relations are a serious problem facing our society.

7. What is the margin of error at that point in the survey?

8. If you decide that you'll only consider your survey a success if the margin of error is at most 4%, it would be perfectly reasonable to want to know how many more people you'd have to survey. Assuming that the result holds at 32%, set up the margin of error formula with a 4% margin of error, a 32% response proportion, and variable n representing the number of people surveyed. (Remember, percentages need to be written in decimal form.

order to find the sample size needed to get that 4% margin of error, we'd need to be able to solve the equation you wrote in Question 8 for n. There are two issues to address: first, the quantity we want to solve for is inside a square root, and second, it's in the denominator of a fraction.

Let's begin by eliminating the radical. To accomplish this, we can square both sides of the equation, because we know that squaring a square root eliminates the radical. But first, it's helpful to rearrange so that the square root is the only thing on one side of the equation.

9. Starting with the equation $0.04 = 2\sqrt{\frac{(0.32)(0.68)}{n}}$ (Aw man, I just gave away the answer to the last question...), divide both sides by 2 to isolate the radical.

10. Now square both sides of the equation.

11. Keeping in mind that the goal is to solve for n, what can we do to BOTH SIDES of the equation in order to get n out of the denominator? Explain what we can do, then, you know, do it.

12. Now you should be able to solve for n, so write how many MORE PEOPLE you'd need to survey to get that 4% margin of error we were shooting for.

13. Now let's go back to the presidential poll we started this lesson with. Use the approval rating you discovered, along with the margin of error, to find the number of people that were polled. Now look again at the description of the poll. How does your result compare to the number of people that were actually surveyed?

14. Suppose that the result of a public opinion poll shows that just 9% of people would be willing to give up their cell phone for 1 week in exchange for $100. How many people would need to be surveyed in order to have a margin of error of ±2% ?

15. Repeat Question 14, but this time for a poll where 50% of respondents say they'd be willing to give up their phone for a week. How does this affect the number of responses needed to get the 2% margin of error?

16. Bonus Question: Look carefully at the margin of error formula. What about it, mathematically, makes your conclusion from Question 15 make perfect sense? Detailed explanation if you want bonus points!

17. Now design your own problem. Find a survey online that interests you, and using the result and the reported margin of error, calculate the number of respondents.

Stats Portfolio Name ______________________

Check each box when you've completed the task. Remember that your instructor will want you to turn in the portfolio pages you create.

Technology

1. □ Find an article on either Gallup.com or Pollster.com that has the term "margin of error" in it, and briefly explain what that means in the context of that particular poll.
2. □ As we saw in this lesson, being able to find the number of people that are needed to get an acceptable margin of error is really useful. Build a spreadsheet like the one below that uses a formula to compute the value of n needed to achieve a certain margin of error given a value of $\hat{p}$. (Hint: You'll need to solve the equation that computes margin of error for n. Questions 9-12 in the Class portion will be a big help.)

	A	B	C
1	p	Margin of error	n
2	0.32	0.01	8704
3	0.32	0.02	2176
4	0.32	0.03	967
5	0.32	0.04	544

Skills

1. □ Include any written work from the online skills assignment along with any notes or questions about this lesson's content.

Applications

1. □ Complete the applications problems.

Reflections

Type a short answer to each question.

1. □ What is the margin of error for a poll? When is it used, and why?
2. □ Describe how being familiar with margin of error might help you in the future.
3. □ Take another look at your answer to Question 0 at the beginning of this lesson. Would you change your answer now that you've completed the lesson? How would you summarize the topic of this lesson now?
4. □ What questions do you have about this lesson?

Looking Ahead

1. □ Read the opening paragraph in the next lesson assigned by your instructor carefully, then answer Question 0 in preparation for that lesson.

Stats Applications Name ______________________________

A stats class was interested in determining the percentage of students on their campus who are opposed to a new policy eliminating all paper books from the campus bookstore in favor of ebooks. After surveying 90 students, the class finds that 33% are opposed to the policy.

1. Calculate the margin of error for this survey. Round to two decimal places.

2. Write a short paragraph describing exactly what your answer to Question 1 tells you about the percentage of the overall student body that is opposed to the policy.

3. The class was unhappy with the margin of error (as well they should), so they decide to survey another 90 students. With the additional data, the percentage of students opposed to the policy goes up to 35%. What's the margin of error now?

4. If surveying more students keeps the percentage at 35%, how many more would need to be surveyed in order to give 95% certainty that the correct percentage for the entire student body is between 32% and 38%?

Photo Credits

Unit 1

p.1: © Creatas/PunchStock RF; p. 3: © Corbis RF; p. 11: © Lars A. Niki RF; p. 19: © Punchstock/Stockbyte; p. 27: © Rich Legg/Getty RF; p. 34: © Tara McDermott; p. 37: © Comstock RF; p. 45: © Design Pics/Craig Tuttle RF; p. 53: © Stockbyte/Punchstock RF; p. 63: © JGI/Blend Images LLC RF; p. 69: © Doug Berry/Corbis RF; p. 72: Courtesy of Park West Gallery/David Najar; p. 76: © McGraw-Hill Education/Ken Kapr photographer; p. 77: © Comstock/Alamy RF; p. 78: © Cat Sobecki; p. 79: © BananaStock/PictureQuest RF

Unit 2

p. 85: © LWA/Dann Tardif/Blend Images LLC RF; p. 87: © Suraj Hemnani–weatherpuppy.com; p. 88 (die): © McGraw-Hill Education; p. 88 (coins): © Brand X Pictures/PunchStock RF; p. 88 (cards): © Cat Sobecki; p. 88 (hand): © McGraw-Hill Education; p. 92: © Photodisc/Getty RF; p. 95 (top): © 1999 New Line Cinemas, *Austin Powers: The Spy Who Shagged Me*; p. 95(bottom): © Jeffrey S. Viano, U.S. Navy; p. 96: © Cat Sobecki; p. 97: © Tara McDermott; p. 99: © 1997 Paramount Pictures, *Titanic*; p. 100: © United States Treasury; p. 102 (top): © Mikael Karlsson RF; p. 102 (middle): © 1984 Columbia Pictures, *Ghostbusters*; p. 102 (bottom): © Andy Crawford/Getty RF; p. 103: © Cat Sobecki; p. 105: © 1985 Universal Picturs, *Back to the Future*; p. 108: © Getty Images/Digital Vision RF; p. 110: © Cat Sobecki; p. 111: © Bruce F. Molnia, U.S. Geological Survey; p. 113: Library of Congress; http://www.loc.gov/pictures/resource/cph.3b46036/; p. 118: © Photographer's Choice/Getty; p. 119: © Brand X Pictures RF; p. 127 (top): © Pixtal/age Fotostock RF; p. 127 (bottom): Courtesy Steve Petteway, Collection of the Supreme Court of the United States; p. 137: © Brand X Pictures/PunchStock RF; p. 147: © Brand X Pictures RF; p. 153: © Wade Clarke

Unit 3

p. 165: © Brand X Pictures/PunchStock RF; p. 167: © Janis Christie / Getty RF; p. 174: © McGraw-Hill Education/Ken Cavanagh Photographer; p. 177 (top): © Glow Images RF; p. 177(bottom): © Stockdisc/Getty RF; p. 180: © The McGraw-Hill Companies, Inc.; p. 181: © Gold Medal Products, Inc.; p. 185: © Don Farrall/Getty RF; p. 191: © Evan-Amos; p. 201: © Brand X Pictures RF; p. 204: © Corbis RF; p. 209: © Cat Sobecki; p. 221: ©REB Images/Blend Images LLC

Unit 4

p. 233: © Ethan Kavet; p. 237: © Cat Sobecki; p. 238: © McGraw-Hill Education/John Flournoy, photographer; p. 245: Courtesy Manette Hall; p. 247: © Dave Sobecki; p. 249: © McGraw-Hill Education; p. 252: John Pallister/USGS; p. 253 (top and bottom): © Dave Sobecki; p. 254: © Joel Gordon; p. 255: © Cat Sobecki; p. 257: Courtesy Mohamed Ibrahim/Clker.com; p. 260: © McGraw-Hill Education; p. 261: © McGraw-Hill Education; p. 263: © Jose Luis Pelaez, Inc./Blend Images/Corbis RF; p. 267: © Ingram Publishing/Alamy; p. 275: © 1965 United Artists, *Thunderball*; p. 285: © Brand X Pictures/PunchStock RF; p. 291: © Cat Sobecki; p. 295: © Jose Luis Pelaez Inc/Blend Images LLC RF; p. 307: © Corbis RF; p. 308: Courtesy Dave Sobecki; p. 319: © Akihiro Sugimoto/age fotostock RF; p. 329: © Dave Sobecki

Index

N

O

P

Q

R

S

T